George Homans Eldridge

A geological Reconnaissance across Idaho

George Homans Eldridge

A geological Reconnaissance across Idaho

ISBN/EAN: 9783337077136

Printed in Europe, USA, Canada, Australia, Japan

Cover: Foto ©ninafisch / pixelio.de

More available books at **www.hansebooks.com**

DEPARTMENT OF THE INTERIOR—U. S. GEOLOGICAL SURVEY
CHARLES D. WALCOTT, DIRECTOR

A GEOLOGICAL RECONNAISSANCE ACROSS IDAHO

BY

GEORGE HOMANS ELDRIDGE

EXTRACT FROM THE SIXTEENTH ANNUAL REPORT OF THE SURVEY, 1894-95
PART II—PAPERS OF AN ECONOMIC CHARACTER

WASHINGTON
GOVERNMENT PRINTING OFFICE
1895

A GEOLOGICAL RECONNAISSANCE ACROSS IDAHO.

BY

GEORGE H. ELDRIDGE.

CONTENTS.

	Page.
Prefatory	217
Topography	217
Drainage system of the Snake River	217
Drainage system of the Columbia River	218
Mountains	219
Canyons and intermontane valleys	220
Glacial action	223
The formations	224
Granites and metamorphic rocks	224
Archean	224
Algonkian	225
Unaltered sedimentary rocks	226
Paleozoic	226
Cenozoic	230
Pleistocene	234
Eruptive rocks	234
General structural features	238
Mining districts	240
Gold and silver	250
Bear Creek district	250
Atlanta district	253
Sheep Mountain district	258
Yellow Jacket district	259
Wood River district	264
Silver City district	271
Placers	273
Coal	274
Salmon City Valley	274
Payette Valley	274
Agriculture	275

ILLUSTRATIONS.

		Page
PLATE	XV. General map of the State of Idaho	216
	XVI. Sketch-map of Atlanta district	254
	XVII. Sketch-map of Wood River district	264
FIG. 38.	Section of Tertiary gravels at Lemhi placer mine	233
39.	Sketch-map of Bear Creek district	251
40.	Sketch-map of Yellow Jacket district	260
41.	Sketch-map of Silver City	272

GENERAL MAP OF

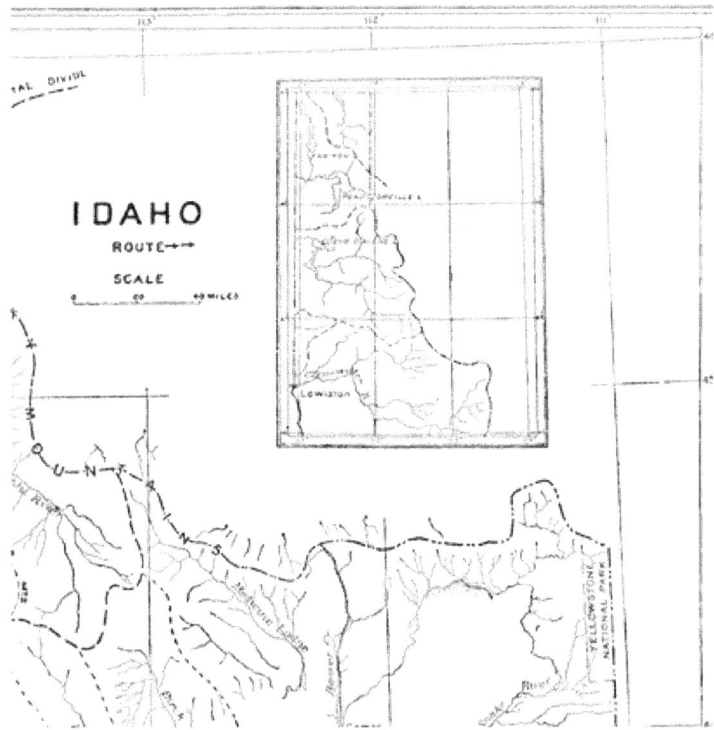

A GEOLOGICAL RECONNAISSANCE ACROSS IDAHO.

By George H. Eldridge.

PREFATORY.

The following pages contain an account of a reconnaissance across Idaho on a northeast line through Boise and Salmon City. The work was authorized as a preliminary to the future geologic study of the State. In its prosecution the available maps have been the atlas sheets of the Survey, between the meridians of 114° and 117° west and the parallels of 43° and 44° 30' north; the maps of the United States General Land Office; and a rough sketch map of Lemhi County.[1] The general map of the State accompanying this report is a compilation from the above sources.

The fossil plant remains referred to in the text have been identified by Mr. Knowlton; the mollusks, by Messrs. Walcott and Stanton; the eruptives, by Mr. Cross; and the chemical analyses have been made in the laboratory of the Survey.

TOPOGRAPHY.

The broad topographic features of Idaho are the drainage systems of the Snake and Columbia rivers, with a vast arid plain along the former stream; a labyrinthine mass of rugged mountains northward from the plain; and a succession of desert ranges on the divide between the Snake River and the Great Basin, along the southern border of the State.

DRAINAGE SYSTEM OF THE SNAKE RIVER.

The Snake Plains, which extend completely across the southern end of Idaho, constitute one of the prominent features of the West. The floor is rolling, consisting of sand and lava underlain by Tertiary sediments several hundred feet thick. These have been cut by the river to a depth of 400 to 1,000 feet, the stream being confined between canyon walls, or lying in a valley with bottom-lands on either side. Following are a few of the determined altitudes of the plains.[2]

---

[1] Prepared by Mr. Birdseye, Salmon City, Idaho.
[2] Chiefly from Gannett's Dictionary of Altitudes, Bull. U. S. Geol. Survey No. 76, 1891.

Altitudes of localities on the Snake Plains.

	Feet.		Feet.
East end of Snake Plains, about	5,000	Shoshone	3,975
Camas	4,822	Bliss	3,269
Market Lake	4,781	Glenns Ferry	2,556
Eagle Rock	4,714	Boise	2,768
Blackfoot	4,505	Nampa	2,489
Pocatello	4,468	Caldwell	2,374
American Falls	4,313	Weiser	2,125
Minidoka	4,287		

These figures show a gentle gradient of approximately 3,000 feet in 400 miles, or an average of 7½ feet to the mile, a little heavier at the eastern end, a little lighter at the western.

Additional elevations along Snake River are: Lewiston, about 1,100 feet; Pasco Junction, Wash., 386 feet; Snake River (mouth), 328 feet. These indicate the maintenance, for the balance of the river's course, from Weiser to the junction with the Columbia, of a gradient of approximately 7 feet.

The drainage of Idaho, with the exception of the southeast corner and a narrow strip of 130 miles at the north, is entirely through the Snake River. The copious waters of this stream are derived from the Continental Divide in the vicinity of the Yellowstone Park, from the great mass of mountains north, and from the divide between it and the Salt Lake and Humboldt regions to the south, just beyond the border of the State. The river forms a third of the western boundary of the State, receiving along this portion the Owyhee, Boise, Payette, Weiser, Salmon, and Clearwater rivers, all draining large areas of country. Of these, the Clearwater rises in the Bitter Root Mountains; the Salmon drains an extensive interior area between the Continental Divide and a parallel range, the Sawtooth, in the middle of the State; the Boise and Payette drain from the latter range west; the Weiser occupies a longitudinal valley parallel with the Snake, from which it is separated by a narrow, rugged range, the continuation of the Seven Devils Mountains; and the Owyhee drains the southwestern corner of the State. The Palouse River, also a tributary of the Snake, drains a small scope of territory just north of the Clearwater, within the western edge of the State.

DRAINAGE SYSTEM OF THE COLUMBIA RIVER.

From the divide north of the Palouse and the Clearwater the drainage is to the Columbia, through the Spokane River, Clarks Fork, and the Kootenai, the first of these, with its tributaries, heading within the State, in the Bitter Root Mountains, the others crossing it from Montana. Of the larger lakes in this system, Cœur d'Alene has an altitude above sea-level of 2,150 feet; Pend d'Oreille, 2,080 feet.

The distribution and relative areas of the several drainage basins are shown on the general map.

MOUNTAINS.

The portion of Idaho lying north of the Snake Plains is an intricate region of lofty mountains and deep canyons. The altitude of the mountains varies between 6,000 and 12,000 feet, while 3,000 to 4,000 feet is a common depth for the canyons. The mountains lie either "en masse"—nearly devoid of topographic system—or in ranges. The development "en masse" is due to early geologic accidents—fracturing, faulting, and folding—and subsequent modification by erosion, the region being one of nearly structureless granite of almost homogeneous texture. The range form, on the contrary, is the effect of erosion either upon pronounced folds in sedimentary beds and foliated granites or upon a complex in which there existed a difference in the texture and hardness of the rocks, decay and disintegration taking place more rapidly in one than in another.

The range of first importance in the topography of Idaho is the Continental Divide. This consists of folded metamorphosed beds—quartzites and schists—the structural and topographic axes coinciding. Second in importance are the ranges dividing the chief drainage basins. Among these are the Sawtooth, in the center of the State, nearly parallel with the Continental Divide; two north and south ranges on the western side of the State—the Seven Devils between the Snake and Weiser rivers, and a shorter ridge between the latter stream and the North Fork of the Payette; a transverse ridge of irregular trend extending from the southern end of the main Sawtooth Range to the Continental Divide, and separating the Salmon waters from those of the Snake to the south; a ridge north of and having the same direction as the last, springing from the Sawtooth in the vicinity of Cape Horn, and constituting the watershed of the main and middle forks of the Salmon; and a range from the Sawtooth west, dividing the Boise and Payette drainage. Of these secondary ranges, the two in the western part of the State are unknown to the writer; the others were crossed once or twice. The Sawtooth and the northern of the transverse ridges are composed chiefly of granite, with local eruptives, while the southern transverse ridge consists in part of granite and in part of more or less metamorphosed sedimentary beds cut by eruptives and folded. Subordinate still to the foregoing are the ranges separating the major water courses within the several drainage basins. Particularly noticeable among these are the Salmon and Lost River ranges and others just to their west, in eastern Idaho, largely composed of folded metamorphic rocks; the north and south ranges between the several forks of the Salmon River, springing from the northern of the transverse ridges and consisting chiefly of structureless granite, except at the east, where quartzites prevail; the ranges between the Boise and Payette basins, also of granite; and finally the Owyhee Range in southwestern Idaho, consisting of granite and eruptives.

The mountains of purest range type occur chiefly in the eastern part of the State, in proximity to the Continental Divide, in a region of uplifted sedimentaries, altered or unaltered quartzites, schists, and limestones. West of this area, in the region of granite, the "en masse" form prevails; but the range type is still present in the Owyhee, Boise, Trinity, Sawtooth, and transverse ranges, where the granite is strongly foliated or the rocks are of different texture and hardness and erosion has removed one class faster than another.

CANYONS AND INTERMONTANE VALLEYS.

Throughout the mountain region are many quite impassable canyons, and others hardly less so. The early prospectors, prior to last year (1894), had made a few passable with trails, and recently the Salmon above Salmon City has been opened by a portion of the State wagon-road. The walls of the canyons are precipitous, sheer drops of 1,000 feet being not infrequent, while the confining slopes rise from 2,000 to 3,000 feet higher. The bottoms are filled with débris, the streams being swift and wild. Among the most rugged of the canyons are: the canyon of the South Fork of the Boise; several along the Salmon; that of the Middle Fork of the Salmon, with its tributary, Loon Creek; and those of the numerous branches of the Clearwater.

Within the mountains, also, are occasionally to be found open valleys from 10 to 20 miles in length by 2 to 7 or 8 in width. They have become the repositories of the products of modern erosion, or of materials derived from the inclosing mountains and laid down in Tertiary times or their early canyons have been filled to depths of several hundred feet with lava-flows, to be recut by later streams to their present levels.

Prominent among these valleys are the following: Along the portion of the Salmon traversed, one at the head of the main fork, east of the Sawtooth Range. There is here a stretch of nearly 40 miles north and south by 6 or 8 east and west occupied by material, chiefly crystalline and eruptive, derived from the surrounding mountains. Long and gentle talus slopes border the valley, while its center is a broad flat, cut to a depth of 5 to 30 feet by the river channel. The valley was probably once the site of a glacial lake, heavy beds of moranial matter being still visible along the base of the Sawtooth.

At Challis is another opening. This in itself is 10 or 15 miles in diameter, but tributary valleys to the east and west carry the general topographic depression some 10 to 15 miles farther back. The Salmon enters the valley 7 miles southeast of Challis through a sharp canyon in crystalline and eruptive rocks, and after a northerly course of about 15 miles, along which are rich bottom and bench lands, again enters the mountains in a canyon continuous—with the exception of a small opening in the vicinity of the Pahsimeri—to within 6 or 7 miles of Salmon City. The valley of Antelope Creek, the tributary from the east, is broad and level, the stream small. The valleys to the west are more rolling, but the creeks, again, carry little water. There is a deep deposit of Quaternary gravel in the Challis Valley, but the evidence of glacial action, along the route traversed at least, is not nearly so pronounced as in the upper valley of the Salmon east of the Sawtooth Range.

The Pahsimeri Valley appears open for fully 15 miles from its mouth. It is 6 or 7 miles broad, with a narrow strip of bottom land along the center, the remainder being bench land underlain with Quaternary debris resting upon crystallines and eruptives.

A third and still broader intermontane valley of the Salmon occurs about Salmon City. This is an illustration of the mountain-locked valleys that have become repositories of materials in Tertiary times. The depth of the pre-Tertiary valley is unknown, but the sediments indicate it to have been several hundred feet greater than at present. The extent of the valley is about 20 miles along the Salmon, in a north and south direction, by 10 miles in an east and west direction. To the southeast it is continued, in the Lemhi Valley, fully 50 miles, maintaining here a width of 5 to 8 miles. A long spur of the Salmon River Range lies in the forks of the valleys. Both the Salmon and Lemhi valleys present gentle slopes from side to center, locally rendered somewhat uneven by the folds into which the Tertiary beds have been thrown. Along the present stream channels, particularly the Salmon, the Tertiary strata frequently form precipitous bluffs 50 to 80 feet high. The first-bottom lands are from a half mile to 1 or 2 miles wide, but there is generally a long, evenly sloping bench on one side or the other which is susceptible of cultivation wherever water is available. The configuration of the valley has perhaps changed slightly from time to time, not only by reason of erosion, but also through the agency of dynamic movements, indicated by flexures in the Tertiary strata.

Of the other intermontane valleys, that of Lost River was traversed only along the two branches forming its head. The northern of these has its source in the Thousand Springs, near Dickey. The valley is here 3 or 4 miles broad and very level; just below it opens to a still greater width and becomes continuous with the main valley below the forks. The west or main fork has a narrow bottom, rarely over a half mile wide, confined between rugged hills from 600 to 2,000 feet high.

The valley of Wood River from a point about 15 miles above Ketchum to its exit from the mountains maintains a width of bottom of between one-half and 1½ miles. The floor is level, and is underlain with a Quaternary wash of varied material. The tributary valleys are usually sharp mountain gorges, their bottoms rarely over one-quarter mile wide.

From Hailey to Boise the route of reconnaissance lay along the southern border of the mountain mass of Idaho; for the eastern half of the distance, in the broad, prairie-like valley of Camas Creek; for the western half—from High Prairie at the head of Camas Creek to the Boise Valley—within the foothills of the mountains. The valley of Camas Creek is from 10 to 15 miles wide, extending directly from the base of the mountains on its north to the low range of hills—partly sedimentary, partly crystalline or eruptive—which divides it from the Snake Plains to the south. The channel of Camas Creek is deep cut, but the lateral streams are rarely depressed more than 10 or 15 feet. Were the water supply sufficient the valley would present some of the most favorable agricultural conditions observed during the reconnaissance.

From the Camas Valley westward for a distance of 25 or 30 miles is a region of high, irregularly eroded hills, extending from the mountains prairieward 10 to 15 miles. The drainage of this area is partly into the Boise, partly direct to the Snake River. The divide between the two waters threads its way with irregular trend through the region, passing finally into the great lava flat which constitutes a part of the Snake Plains at their western end. From this divide, at the head of Indian Creek, a high range passes northwest and walls the Boise intermontane drainage basin from the general valley without. The Boise River debouches from a rugged defile in the range at a point about 12 miles southeast of Boise City. In the vicinity of Little Camas Prairie, however, the early topographic rim, confining on the south the intermontane drainage of the South Fork of the Boise, has been deeply degraded, and but for the cutting of the stream—its recutting, even, through the comparatively recent lavas—at a more rapid rate than the denudation of the rim, the waters of the South Fork would have found an exit at this point direct to the Snake. In the hill area just described the valley of greatest extent is that of the Little Camas, 7 or 8 miles long and from 1 to 5 miles wide. Meadow lands occupy its upper, narrower portion, while below it assumes a prairie aspect and is underlain by a broad flow of lava. A ridge 2,000 feet high extends across the southern end of the valley, and continues eastward for several miles as the southern border, also, of the Big Camas Valley. The entire foothill region between the head of Big Camas Creek and the main Boise Valley west of the mountains is one of granite, lava, and other eruptives in dike form, and the effect of such varied rock assemblage has been a most irregular development of topographic features. The lava generally occurs as a thin sheet or a succession of thin sheets upon the granite, and wherever it is present the country becomes rough and difficult to travel.

North of the region just described and within the rim of the Boise drainage basin is a flat, 8 to 10 miles in diameter, known as Smiths Prairie. The floor is wholly of lava, which extends, also, for some distance into the tributary valleys. The surface is locally smooth, rugose, or gently undulating, and in only one or two instances is it relieved by low hills. The general altitude of the surface is about 4,800 feet, a little lower than the small lava prairie a few miles to the southeast at the mouth of Little Camas Creek. Along the southern edge of the prairie the South Fork of the Boise has cut a gorge over 1,000 feet deep, in the precipitous walls of which appear two seemingly distinct lava-flows, one occupying the upper 500 to 700 feet of the gorge, the other 200 to 300 feet at the bottom. The relative ages of these flows were not determined, but either might be the older. The flows may be traced at intervals the entire length of the South Fork, and also along the main stream below, to its debouchment from the mountains. At the mouth of Moores Creek they are joined by others which extend nearly to Idaho City. Usually the lava forms narrow ribbons adhering to the sides of the canyons. These appear to drop in altitude as the canyon is descended, indicating an ancient lava river down an earlier-eroded gorge, with occasional lava lakes in the openings.

Twenty-eight miles north of Boise, at Horseshoe Bend, on the Payette, is a small mountain-locked valley carrying deposits of Tertiary age. The length of the valley is 10 to 12 miles, the width 4 or 5 miles, the trend northeast. Along the western edge the Payette flows for a distance of 5 or 6 miles, entering and leaving by a sharp, rugged canyon in the granite of the Boise Range. The valley is apparently one of erosion. Its configuration is somewhat irregular, both upon original outlines and from the later encroachment of the heavy talus slopes along the base of the mountains. Moreover, the underlying Tertiary beds here and there project through the Quaternary in buttes or ridges. Gentle folding has taken place here, as in the valley at Salmon City.

One of the most peculiar valleys encountered, of which the topographic origin was not determined, is that locally known as "Prairie Basin," about 45 miles southwest of Leesburg. This is a high intermontane depression, apparently of considerable depth originally, but since filled by heavy deposits of glacial drift, which in still later times has been partially removed through the channels of Big Creek and its tributaries. The valley surface is now in long, rounded ridges and hillocks. Owing to its altitude, to the extent of opening, and to the heavy growth of grass covering its every acre, the "basin" is easily recognized from distant peaks.

GLACIAL ACTION.

Evidences of early glaciers were observed at many points along the route of reconnaissance, and it is probable that their existence was quite general throughout the mountain region of Idaho. Conspicuous localities in the southern half of the State are the Trinity and Sawtooth ranges. In the upper portions of many of the valleys heading in the former are glacial bowlders, grooved rock surfaces, roches moutonnées, and lateral and terminal moraines. North of the Trinity Range, and also east of the Sawtooth, are numerous glacial lakes, 1 to 2 miles in diameter, their waters held back by terminal moraines. East of the Sawtooth the mass of débris from the earlier glaciers is enormous, extending 50 or 60 miles along its front and forming a belt from 5 to 8 miles broad. This is cut by the streams of the present day, and now forms a rough, often heavily timbered slope most difficult to traverse. Other observed localities are the ranges west of Salmon City, particularly about the head of Panther Creek and the several tributaries of the Yellow Jacket, a branch of Camas Creek. Prairie Basin has already been noted as the possible site of an ancient glacial lake, the heavy deposit of bowlders having been derived from the surrounding mountains. Indeed, nearly all of the valleys traversed in the reconnaissance show, at their heads, more or less evidence of former glaciers. Even now ice action is rife at the higher altitudes, where the snows are exceedingly deep and their disappearance is rarely complete from one season to another.

THE FORMATIONS.

The rocks occurring in the southern half of Idaho embrace granites, gneisses, syenites, schists, quartzites, limestones, calcareous and non-calcareous shales, sandstones, clays, and eruptives in great variety. The granites and syenites, in part at least, are probably of the Archean age; the schists, Algonkian. The quartzites are distributed from Algonkian to Carboniferous, while the limestones may include both Silurian and Carboniferous. The age of the great calcareous shale series of the Wood River and neighboring districts is undetermined, but the evidence points to the Carboniferous. It is undoubtedly Paleozoic. The sandstones and clays encountered are all of Tertiary age, Eocene (?) and Neocene. Post-Pliocene gravels are abundant. The eruptives are possibly of all ages, from Archean to Recent. The assignment of the several series of rocks, except the sub-Carboniferous and Tertiaries, is provisional, being without fossil evidence; lithology and stratigraphy alone form the basis of reference. Fossils might be found at several horizons, but the exigencies of the trip did not permit careful search.

GRANITES AND METAMORPHIC ROCKS.

ARCHEAN.

To this group are provisionally assigned the granite and gneiss, but there are instances where, by reason of included bands of calcareo-micaceous or quartzitic slates, this reference to the Archean instead of the Algonkian is questionable. Again, it is probable that, in part at least, the granite is of igneous origin.

The granite is of wide occurrence, and, in the main, of a single type, with three or four regional modifications. The type rock is gray, moderately coarse in texture, and composed of feldspar, quartz, and mica, with few accessory minerals. The feldspar is white, chiefly orthoclase, with perhaps an occasional small amount of plagioclase. The orthoclase is often porphyritically developed, while the color varies locally, but rarely, to a faint pink. The quartz is generally granular. The mica includes both the black and white varieties, the former predominating, but showing considerable variation in amount. It is distributed irregularly throughout the mass of the rock, or is lineally disposed, when it imparts a more or less definite foliation. An accessory mineral is hornblende, in fine and coarse crystals, but on the line of reconnaissance its presence is local and somewhat rare. It occurred notably in the granite of Trinity Range, 8 or 9 miles west of Rocky Bar, and again in conspicuously large crystals along the lower portion of Napias Creek, 20 to 30 miles west of Salmon City. In the latter locality a further modification of the type granite takes place in the development of the orthoclase crystals to an extraordinary size—from 1 to 3 or 4 inches in the direction of the vertical axis. The outlines of the crystals are usually somewhat rounded. On exposed rock surfaces the crystals weather in knobs, but fresh fractures extend through rock and crystals alike, often along cleavage planes of the latter, the rock then presenting the appearance of a more or less uniformly crystalline granular matrix sharply relieved by the smooth, brilliant faces of the porphyritically developed feldspar. To the granite of this description the designation "bird's-eye" may be appropriately applied.

A second departure from the type granite occurs in the precipitous walls of the Loon Creek Canyon, 3 or 4 miles below Oro Grande. Here the gray variety is wholly replaced by one of deep pink, derived from orthoclase of this color. A faint greenish tint is sometimes induced by the decomposition of the biotite. The texture of this granite is fine to coarse, and the grain even. The areal extent is undetermined, but it outcrops at intervals for several miles.

A third and local modification of the normal granite takes place in proximity to mineral veins. This consists in the loss of a very large proportion of the mica, the remaining feldspar and quartz constituting a somewhat conspicuous rock, by means of which the course of the vein may readily be traced, and which, in fact, may enter into the composition of the vein itself as "ledge matter." The clearest illustration of this occurrence is to be seen in the mining camps in the vicinity of Rocky Bar and Atlanta.

Typical gneiss occurs in some of the spurs of the Sawtooth Range, notably about the drainage system of Upper Redfish Lake. Wherever observed it has the mineral composition of the typical gray granite. In the locality referred to, the foliation is so complete as to create the impression of bedded strata, while the effect upon topographic lines is markedly that of an unaltered stratified rock.

The granites and gneisses prevail in the mountains of the western half of the State, while in the eastern they occur as range cores in connection with schists, quartzites, or limestones. None of the latter rocks, however, were encountered in the western half of the State.

ALGONKIAN.

To this is provisionally assigned the great series of micaceous, quartzitic, and chloritic schists of eastern Idaho. The reference is based merely upon lithological character and a resemblance to other beds in the Cordilleras which have already been so assigned. The series embraces, together with the schists, numerous beds of quartzite, and all have the general field appearance of clastic rocks. Many of the layers contain a very considerable amount of carbonate of lime. The series in regions of strong development has a probable thickness of 3,000 to 4,000 feet, and is believed to be unconformable with the granite. In any event there was a time-break prior to the deposition upon the granite of the superincumbent strata, for the same series does not everywhere follow in the areas brought under observation.

The region of crystalline schists is distinctively the eastern half of the State, although in this portion are found also the Paleozoic measures, as well as the older granites forming cores of many of the ranges. The strongest development of the series observed on the route of reconnaissance was in the Continental Divide east of Salmon City, in the region of Big Creek and its tributaries, and in the range separating the drainage of this stream from that of the main Salmon. The series is also exposed at several points along the Salmon above Salmon City, notably, from a point about 9 miles above the city to one 18 miles above. It is then cut out by eruptives of undetermined extent; after a partially covered stretch of 10 miles it again appears along the river for a distance of about $2\frac{1}{2}$ miles; eruptives again succeed, the schists finally disappearing in a small, disconnected outcrop near the mouth of the Pahsimeri.

Schistose rocks occur associated with the slates of Lost and Wood rivers, but it is somewhat doubtful if they belong, in their entirety at least, to the great series of crystalline strata just described. They are here apparently associated more closely with Paleozoic measures.

UNALTERED SEDIMENTARY ROCKS.

PALEOZOIC.

The succession of beds in southern Idaho is difficult of determination, owing, first, to a marked difference in their petrographic characters from recognized formations elsewhere in the Rocky Mountains, this preventing their assignment to a definite position in the scale of formations; secondly, to frequent interruptions of continuity by eruptives; and in the third place, to folds and faults.

Later than the granites, and probably also than the schists described above, is a great body of pink and white quartzites, of at least 1,500 feet maximum thickness. They are heavy bedded, hard, and uniform in texture and composition. Their greatest development is along the Salmon River from a point about 5 or 6 miles below the mouth of the Pahsimeri to within 6 or 7 miles of Challis. At the lower end of this stretch they are thrown into a prominent anticline, the eastern end of which is cut by the sharp gorge of the river, showing them in uninterrupted succession for 1,000 to 1,500 feet. Numerous other similarly disposed anticlines occur between this point and Challis, all in these measures. In none was Archean observed, and none was capped by higher strata of different nature. A small exposure of mica and quartzitic schists (Algonkian?) occurs just below the mouth of the Pahsimeri, and although of the same strike as the quartzites on either side, they are separated from them by intervals of 3 miles or over occupied by eruptives or a Quaternary wash, and the relations between the two series are thus obscured. On Deer Creek, about 6 miles above its confluence with Wood River, is a local body of quartzite, which somewhat resembles those just described along the Salmon. It rests directly upon Archean granite, and is overlain by the dark-blue and black calcareous slates which form so important a feature of the Wood River district. The areal extent of this quartzite was not determined, but it is apparently small. A little south of Deer Creek it disappears with marked rapidity, the slates coming down on the granite. Similar quartzites probably exist in large bodies in the Salmon River Range also.

This quartzite is evidently to be classed with the older sedimentary rocks encountered; moreover, it closely resembles the established Cambrian quartzite of Colorado; on these grounds it is here referred to the Cambrian.

In the high range of mountains forming to the south the watershed of Wood River and to the north and northeast that of Salmon and Lost rivers are several thousand feet of quartzites, slates, conglomerates, calcareous shales, and limestones, which it has been impossible in the time available either to segregate into formations or to refer to definite ages. The range itself is the southeast extension of the Sawtooth, and carries some of the loftiest peaks in Idaho, notable among them being Mount Hyndman, 12,000 feet, 10 to 15 miles southeast of Ketchum. The range is an anticline, the core being gray granite, of the same type as that of the main Sawtooth, Trinity, and Boise ranges, and with a width of exposure along the Ketchum-Challis road of between 2 and 3 miles. Resting upon the granite on both the north and south sides of the range is the series of sedimentary beds mentioned above. The stratigraphic succession of the several members of the series is not confidently determined, and, moreover, there is an apparent difference in the succession on the two sides of the range.

On the north side, about the heads of the several branches of Lost River, quartzites and slates, dipping northeasterly, follow one another in quick succession. Black limestones are also present, and in some localities, not visited, they are apparently of considerable importance, judging from bowlders encountered in the valley. It is doubtful, however, if any single layer is more than a few feet thick. For 2 miles above East Fork of Lost River, large bodies of lava occupy both valley and hillsides, interrupting the series of sedimentary beds. Below East Fork, however, the quartzites and limestones, or at least quartzites with strongly calcareous layers included, reappear, continuing for 8 or 9 miles down the valley to the point where the road turns from the river northward toward Dickey. Within this distance the quartzites perhaps predominate, but black and probably often calcareous slates are not infrequent. The calcareous nature of some of the quartzite layers is particularly marked a short distance below East Fork. For the lower 2 miles of the distance there occur in the series a number of conglomerate layers 1 to 10 feet thick. These are composed of a mass of cherty, subangular pebbles, and in the heavier beds, of quartzite and limestone débris in addition, the whole cemented by fine, granular material of the same nature. East of the Dickey road, after its turn northward from the valley of Lost River, what are probably sub-Carboniferous limestones appear in a prominent hill, a spur of the high range to the south. This limestone, which overlies the quartzite and slate series, outcrops in great bodies in the mountains to the north, forming the western periphery of the Thousand Springs Valley.

South of the Sawtooth Range, along Trail Creek and the Ketchum road, the succession of beds is somewhat different from that on the north side. Dark quartzites prevail immediately above the granite, followed after several hundred feet by a thick zone of light-colored, white and gray quartzites. About 6 miles from the summit the entire series is repeated, though whether in natural succession or by faulting was not ascertained. It is then overlain by several hundred feet of dark-gray and black slates or shales, probably often calcareous, which continue nearly to the mouth of Trail Creek Canyon. At this point the succession is interrupted by eruptives, which extend for a considerable distance along the Wood River Valley on its eastern side. West of Wood River, opposite Ketchum, is a heavy body of limestone, somewhat resembling in its massive character, its mode of weathering, and its general appearance the sub-Carboniferous of the Rocky Mountains, but the age was not definitely determined. This limestone is called by the miners the "gray limestone," in distinction from the "blue limestone," which occurs as thin beds in the series of shale overlying. The dip is here down the river (southward). The overlying shales are gray, dark-blue, and black, and carry numerous one-foot to six-foot limestone bands distributed through them. The thickness of this shaly series is estimated at 5,000 feet. It constitutes the principal ore-bearing series of the Wood River district, though the lower slates and quartzites occasionally carry mineral.

On Deer Creek, a tributary of Wood River from the west, an eruptive again cuts into the sedimentary beds and occupies the hills to the north of the valley for a mile above its mouth. Above this, however, quartzites come in, followed by a few hundred feet of heavy-bedded dark-blue limestones, which bear considerable resemblance, in texture, weathering, and general appearance, to the sub-Carboniferous. Black calcareous slates of the general character of those already described follow the limestones, and are in turn succeeded by granite, just above Warm Springs, 4 miles above the mouth of the creek. This succession of shales by granite is possibly due to faulting. The granite extends up the valley for about 3 miles, when 400 or 500 feet of the heavy-bedded

Cambrian-like quartzite succeed, overlain by blue and black slates and limestones similar to those noted farther down the creek. The quartzites of this second area of sedimentary beds did not appear in the first series, east of the granite at the Warm Springs. The dip of the series below the Warm Springs is doubtful; west of the granite block, however, it is westward, bending around gradually to the southwest and south at the head of Deer Creek and its tributary from the south, Red Cloud Gulch. The black shales and limestones of Deer Creek are undoubtedly a part of the general series which constitutes the leading formation of the Wood River district, but their horizon in the latter is undetermined. South of Deer Creek, east of Red Cloud and Narrow Gauge gulches, the blue calcareous slates are in direct contact with the granites to the east, the quartzite observed in the valley of Deer Creek having here disappeared. Whether faulting or non-deposition is the cause of irregularity in this succession of the beds is unknown.

It is possible that the foregoing series of rocks from the granite up, perhaps 10,000 to 15,000 feet in all, fall wholly within the Paleozoic—Cambrian and younger systems; but except in the case of the sub-Carboniferous and overlying shales, neither their age, division lines, nor intersuccession was satisfactorily determined. The heavy bed of white and pink quartzite on Deer Creek belongs, perhaps, to the Cambrian; the dark quartzite series at the head of Trail Creek, and the calcareous slates, quartzites, and conglomerates of Lost River—from an unknown member of which the one or two fossils collected have been determined by Mr. C. D. Walcott to be not older than the Trenton—are apparently of a post-Cambrian age; the sub-Carboniferous is recognized; but the overlying shales are, again, beyond a general reference to the Carboniferous, in doubt. Lithologically they bear a slight resemblance to the Weber of Colorado and Utah. In the entire Wood River district there is much and varied faulting as well as folding, and only work in great detail will effect a solution of the many geological problems presented.

The foregoing series doubtless occurs in many ranges to the east and northeast of the localities described, especially in the mountains east of the Thousand Springs Valley and of Lost River.

The sub-Carboniferous, which, of the Paleozoic series is most clearly recognized, is, excepting possibly in the Wood River region, represented by the massive blue cave limestone, so characteristic of it throughout the Rocky Mountain region. Its identification is based upon lithological resemblances and the contained fossils, although no collection of fossils was attempted. The thickness is between 100 and 400 feet. The limestone occurs in especial force in the several ranges between the Challis Valley and the main fork of Lost River, but it is said to be present in many localities in eastern Idaho. Its relations to underlying beds were unobserved except in a single locality, in the forks of Lost River, where it apparently succeeds the series of calcareous quartzites, slates, and conglomerates described above. On the divide between Antelope Creek and Lost River it is overlain by drab calcareous shales resembling those in the Wood River district, though possibly less metamorphosed.

In the range of mountains forming the southwest side of the Warm Spring and Antelope valleys, just east of the Salmon River, are beds, possibly 500 to 600 feet in total thickness, which are partly quartzite, partly limestone, the latter pinkish-drab, massive, and somewhat resembling the Silurian limestones of the Colorado areas. In the hurried examination given them, however, no fossils were found. In this locality, also, on the lower flanks of the range, are numerous large hot-spring deposits, the springs being now extinct. At the eastern end of the range, overlying what from a distance is apparently the sub-Carboniferous limestone, are about 50 feet of red beds of unknown composition. Above these are several hundred feet of calcareous shales, the same as those already mentioned on the divide between Antelope Creek and Lost River. The Little Lost River Range east of the Antelope and Thousand Springs valleys is reported by prospectors as being composed of the sub-Carboniferous limestone and overlying shales, and from a distance this seems to be the case.

CENOZOIC.

This system is represented in Idaho by the great series of sedimentary beds underlying the Snake Valley and by others which occupy certain of the intermontane valleys. Paleobotanic evidence points to the Eocene or Miocene as the age of the intermontane sediments, and molluscan and mammalian remains to the Pliocene as that of the Snake River beds.

The materials of the Snake River beds were derived largely from the mountains to the north and south of the valley, though doubtless more or less detritus from the region of the headwaters has been commingled with the other sediments for its entire length. The material, in the western half of the State at least, is chiefly of granitic origin, consisting of quartz and feldspar grains, often with a kaolin-looking cement, slightly ferruginous. The prevailing rock is a fine to medium grained, friable, gray sandstone, but clays occur, and also conglomerates. The formation was examined only locally, and there are doubtless many variations.

Associated with the Snake River beds, and in some instances interstratified with them, are flows of the lava which in one locality or another is such a well-known feature of the Snake Plains. Apparently the outpouring of this rock took place, in part at least, while the sedimentary beds were still being deposited. The sedimentaries outcrop along the Snake River in bluffs from one hundred to several hundred feet high, and also occur in benches of considerable elevation next to the mountains both north and south.

The age of the Snake River beds is probably Pliocene. *Melania taylori* Gabb and *Lithasia antiqua* Gabb, together with an undetermined vertebra (carniverous), have been found by Mr. Arthur Foote, at Glenns Ferry, on the Snake, in Elmore County, and Prof. O. C. Marsh personally reports Pliocene mammalia from the same series of beds on the north side of the Boise River, a few miles below Boise.[1]

The intermontane valley of the Salmon and Lemhi rivers, in the center of which Salmon City is situated, is occupied by a considerable thickness of Tertiary beds, the materials for which were derived from the early formations of the inclosing ranges. These materials are, for the most part, of granitic or quartzitic debris, and the beds are either clays, sandstones, or conglomerates.

The clays occupy a large area in the center of the valley. They are light-green and red, forming a conspicuous feature in the landscape. The colors are in two zones, the red at the higher level, nearer the periphery of the valley, and possibly a coloration of later times. The clays are slightly arenaceous in some layers, in others distinctly sandy, carrying even occasional thin bands of sandstone or conglomerate near the top. This conglomerate is a mass of small, round, or lenticular pebbles of half-inch maximum diameter, chiefly quartzite. Near the same horizon are thin layers of hard, white or brown, homogeneous clay, which usually has an abundance of minutely divided vegetable matter through it, sometimes sufficient to render it lignitic. Here and there fragmentary stumps of trees are found in a carbosilicified condition.

The sandstones of this intermontane Tertiary series occur at the base of the measures exposed, just beneath the clay division, and in all are probably 400 or 500 feet thick. They outcrop in bold bluffs along the Salmon. Their material is chiefly quartz. They are light yellow or gray and for the most part massive, though in some layers thin-bedded and even shaly. The gray beds are usually the finer in texture, the yellow sometimes approaching a grit. The series carries plant remains, leaves, stems, etc., which locally are in sufficient quantity to form with the shale a lignitic band. In one instance, just below Salmon City, this carbonaceous matter is distributed through 4 or 5 feet of shales, forming a coaly slate, within which is a 6-inch bed of dark-brown, woody lignite. So far as at present exposed, this is not of economic value. Overlying the slate are 10 feet of sandstone, succeeded by 5 or 10 feet more of a very hard, moderately coarse quartzite-conglomerate, which is somewhat ferruginous, and often fractures across matrix and pebbles alike. Within a short distance of this outcrop is another exposure of a similar conglomerate, in appearance 30 or 40 feet higher up than the first, but not definitely so determined. This second conglomerate in particular bears a resemblance

---

[1] See also various references in Bull. U. S. Geol. Survey No. 84, Correlation papers—Neocene, by Dall and Harris, pp. 285 and 286, 1892.

to certain beds along the base of the Continental Divide in the vicinity of the Lemhi placer mine on Kirtley Creek. Just beneath this conglomerate bed, along the Salmon, are some white sandstones, more or less argillaceous and ferruginous, carrying kidney-shaped iron concretions, and leaf-bearing. Among the forms collected Mr. Knowlton has recognized—

> *Sequoia langsdorfii* (Brong.) Ung.
> *Glyptostrobus europæus* Brong.
> *Equisetum* (?) sp.
> Dicotyledons: *Ficus* (?), *Quercus* (?), etc.
> Plant stems.

Mr. Knowlton adds the following remarks in the letter submitting the results of his examination:

Neither of the conifers can be relied upon to prove close questions of age, for they have a considerable vertical range. *Glyptostrobus europæus* is, however, absolutely, and *Sequoia langsdorfii* almost exclusively, confined to the Tertiary. Both species have a wide geographical range, and are very abundant forms. The first is abundant in this country in the Fort Union group, Upper Laramie (Fort Union) of Canada, Mackenzie River, Alaska, and Arctic Miocene in general. In Europe it is mainly confined to the Eocene and Miocene. *Sequoia langsdorfii* has been once reported from the true Laramie, but I regard this identification as extremely doubtful. It is very abundant in the Fort Union group, and is also found at the same places as the other. The Dicotyledons are fragmentary but seem to belong to such genera as Ficus, Quercus, etc. The Equisetum and vegetable stems are worthless for stratigraphic purposes.

The evidence, incomplete as it is, shows the Tertiary age of these beds, but whether they are Eocene or Miocene it is impossible to say.

In the bluffs of Kirtley Creek, about 1½ miles below the Lemhi Placer Company's bar, is the following exposure:

	Feet.
At top of bluff, obscured beneath Pleistocene gravel	20
Argillaceous sands	5
Pure sands, concretions at base	10
Coarse gravel	10
Bright-red clays, very pure	5
Red clays with pebbles finer than in the gravel above	5

Gravel probably still underlies, but it is covered.

The position of the beds is nearly horizontal, with a possible slight dip to west. Their relation to the other beds in the valley could not at the time be determined, but they are quite likely younger than the series of green clays, at one point, indeed, apparently resting upon them or a nearly allied stratum. Moreover, apparently the red clays everywhere occur at a higher altitude than the green.

The conglomerates of this series of Tertiary beds outcrop in great force about the periphery of the Salmon City Valley, especially at the west base of the Continental Divide, and are difficult of reference. Being auriferous, they are the most important, from an economic standpoint, of all the beds. They are exposed in the cuts of the Lemhi

Placer Mining Company on Kirtley Creek, about 2 miles from the base of the range. At this point the section given below was obtained.

The gravels of this section, Pleistocene and Tertiary, are made up of quartzites and schists derived from the neighboring mountains. The two are readily distinguished, the Pleistocene being loose, easily disintegrated, and carrying in the interstices of the bowlders and pebbles considerable earthy matter; the Tertiary gravels being a compact mass of large pebbles, up to 1 foot in diameter, in a grit matrix, the whole moderately hard and difficult to hydraulic.

The unconformity between beds c and d is evidenced not only by the wavy lines between the two, as shown in the sketch, but also by the fact that the gravel beds c rests successively upon several of the underlying beds. Both series of strata, however, those above and those below the line of unconformity, are alike flexed in the crumpling that has taken place in the region. The beds d–f are considered "bed-rock" by the

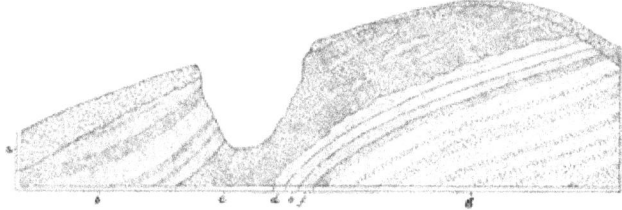

FIG. 28.—Section of Tertiary gravels at Lemhi placer mine.

a, Pleistocene gravel, overlying unconformably the Tertiary series, 9-10 feet; auriferous. b, Series of sandstones with some gravelly layers; average thickness, 30 feet. c, Coarse conglomerate; material, quartzite; 20 feet; unconformable with d; auriferous. d, Yellow sandstone, solid bed, 3 feet; leaf bearing. e, White sandstone, solid bed 5 feet; leaf bearing. f, Very white sandstone, thin bedded, fine grained; 5 feet. (d–f, Bed-rock, of miners.) g, Series of sandstones and conglomerates in equal proportions; 35 feet shown, but it extends below the present depth of the cut; auriferous.

miners, though the series below is said to be auriferous, even in paying degree under favorable conditions of water and work. The thickness of the conglomerate series is unknown; it possibly reaches 300 or 400 feet.

The Pleistocene cap to the series is, in part, probably, wash direct from the mountains, and in part, doubtless, the product of subaerial denudation of the Tertiaries.

On the hills of greater height on both sides of Kirtley Creek, at an altitude of about 200 feet above the stream, is a bed of coarse quartzite-conglomerate, very hard, fracturing across matrix and pebbles alike. Its relation to the other beds was undetermined.

It was impossible, within the short time available, to determine completely the relative stratigraphic position of the three or four varieties of Tertiary rocks described. The sandstones with their subordinate conglomerate beds, which outcrop in the bluffs of the Salmon River, are beyond question overlain by the green clays, and these in turn, perhaps, by the bright-red clays and associated gravels outcropping in the bench lands north of Kirtley Creek, below the placer mines. Concerning the position of the conglomerate series at the mines, however, there is more doubt, for the strata underlying the area to the west were covered for a considerable distance by wash. The dip of the conglomerates at the mines is 15° to 50° westward, and they pass beneath the general surface of the long, gentle valley slope. If they are not brought up by flexures within a mile or two below the Lemhi placers, they are succeeded more or less directly by the green clays, and would therefore, in this case, correspond to the sandstone-conglomerate series along the river bluffs. Certain resemblances between the two have already been noticed. It is possible that they may wholly underlie the river outcrops, the latter sandstones perhaps being concealed nearer the border of the basin in the intervals of no outcrop.

The valley of the Payette River, at Horseshoe Bend, also carries a series of intermontane Tertiary beds, the materials for which were doubtless derived from the inclosing granite ranges. The series is in general a succession of sandstones with some clays and conglomerates, carrying plant remains which in some instances have been sufficiently abundant to form carbonaceous shales, in others coal itself. In appearance the Payette beds resemble those at Salmon City; in composition and their place of derivation they may be more closely affiliated with the Snake River series. Of their age, however, except that they are Tertiary, no evidence was obtained.

PLEISTOCENE.

This consists of gravels occurring along stream bottoms, of the products of subaerial disintegration, or of morainal débris. These deposits will not be discussed beyond mention in connection with other points in the report.

ERUPTIVE ROCKS.

The eruptive rocks of southern Idaho present a great variety of types, and their manner of occurrence ranges from a simple dike to a series of sheets surrounding a probable volcanic center or neck. The flows have probably taken place at many stages in the development of the region, from Archean, possibly, to late Tertiary, and even at the present day traces of volcanic activity remain in innumerable hot springs throughout the State.

Along the front of the Boise Range are eruptives of a number of types in dikes, irregular masses, and sheets.

Among the dikes and irregular masses is a boss of andesite in the bluffs on the north side of the Boise Valley, a mile above Boise. Prior to the erosion of the bluffs this andesite was overlain by the Tertiary sediments, in which there was a slight upward bowing. A portion of the andesite has been identified as of the augitic variety; it is dark-gray and banded. The associated variety is light pink-gray and vesicular. The groundmass is, possibly, a devitrified glass, throughout which are small crystals of decomposing triclinic feldspar, and some magnetite. The interrelation of the two andesites was not determined.

About 5 miles east of Boise, in the granite of the Boise Range, near the edge of a prominent basalt flow, in the Flannigan prospect tunnels, are several narrow dikes of aphanitic syenite (lamprophyre), having a general trend of N. 50° to 70° W., with a dip of 45° to 80° SW., apparently coincident with the foliation of the granite. A vein of quartz with a slight admixture of granitic material lies between two of the dikes, at a little distance from each, and is said to be gold bearing.

On the Boise Idaho City stage road, from 2 to 3 miles above Boise, and just at the edge of the granite mountain slope, is an outcrop of rhyolites, of an area somewhat less than a square mile. The variety of rock most typical of the locality is, perhaps, a microspherulitic rhyolite, light-gray to faint pink, with small porphyritic crystals of sanidine and some quartz. This alternates with a mottled variety, pink and dark gray, with the same porphyritic development of crystals. Occasionally the groundmass is microcrystalline or cryptocrystalline, but it is more frequently glassy. Perlitic, spherulitic, and vesicular varieties also occur. Fluidal structure is common throughout the entire mass. The rhyolite cuts the gray granite of the Boise Range, forming an apparently irregular body, the maximum dimension having a northwest-southeast direction.

About 6 miles northeast of Boise, and north of the Boise-Idaho City road, in the slopes of the range, is a dike of quartz-porphyry running through the granite in a general northwest-southeast direction. The dike is not over 3 or 4 feet wide, and with the granite on either side has been prospected for gold.

A hundred yards west of the above outcrop, and also at two or three other points on the lower slopes of the Boise Range, notably one a little south of the Boise-Idaho City road, about 2 miles from Boise, are deposits of a brown, porous, stratified, rhyolitic tuff, carrying a large amount of angular and subangular granitic débris, in fragments up to 1 or 2 feet in diameter. The deposit occurs at various altitudes, but its relations to the Tertiary beds of the region are not clearly shown.

One of the most widely distributed of the eruptives is basalt. This forms a large part of the floor of the Snake Plains, and occurs in isolated sheets, bosses, and dikes at many points in the mountains. Prominent among these are two or three independent sheets on the western slopes of the Boise Range; an enormous flow along the South Fork of the Boise River within the mountains; others, very heavy, above the three forks of Camas Creek, a tributary of the Middle Fork of the Salmon; and several along the Salmon River between Salmon City and Challis. Of those on the western slopes of the Boise Range a typical occurrence is a remnant, a half mile in area, about 5 miles east of Boise, 1.000 or 1,500 feet above the valley. The flow filled the early gulches and capped the intervening ridges, its surface acquiring a position of rest but slightly at variance with the general slope of the range. In later times the gulches have been recut to the granite beneath. Beyond the present limits of the flow no trace of connecting bed of basalt could be found; it was probably, therefore, a small isolated body, welling up through a local dike.

Between Slaters Creek and the Boise the basalt constitutes the floor of a high mesa which extends for 10 to 15 miles along the base of the mountains, with the appearance of having at one time been continuous with the lava plains of the Snake Valley. The mesa has a gentle slope of 2° or 3° to the southwest, the basalt resting directly upon granite. In the gorge of the Boise the basalt forms, for 4 or 5 miles from the mountains, sharp vertical walls, the columnar structure being well developed on both upper and under surfaces of the flow. The thickness of the sheet varies from point to point, averaging perhaps 75 or 100 feet.

Within the mountains the basalt follows the main Boise to the South Fork, passing up this nearly to Pine Grove, a few miles northeast of Little Camas Prairie. It is also reported on Moores Creek nearly to Idaho City. None is said to occur on the main stream above the mouth of South Fork. There appear to have been at least two distinct periods of flow, the remnants of which are attached to the canyon walls, and lie, one, 500 to 1,000 feet above the present stream level, the other but 50 to 200 feet above. Which is the older was not determined. Whether it would be possible, by descending the canyon from Pine Grove, to trace the respective flows continuously, or at least to trace a particular flow entirely through, can not be stated, but it is a peculiar feature of their occurrence that the different benches of basalt are everywhere found practically at about the same height above the present water level; they fall with the stream, impelling the belief that the lava once flowed through the canyon as does the stream of to day.

Smith and Little Camas prairies are two broad intermontane flats on the course of the South Fork, the sites of ancient valleys refilled to their present levels with basalt of the same flows as in the canyon. The canyon is recut to a depth of 1,000 to 1,500 feet, presenting cross-sections of the valleys, which, as regards the basalt, are the same as those of the canyon in general. The precise nature of the two great lava lakes is unknown. They are evidently near the source of the basalt, and the valleys filled by them seem to have been the receiving reservoirs, in part for a stream from a source farther up the Boise, in part for flows from fissures in some of the side valleys tributary to the basins, and, perhaps, at the outset, from fissures in the bottoms of the valleys themselves. From the lakes the flows continued down the early Boise Canyon to the open valley of the Snake, receiving from the tributaries on either side greater or less accessions. This was eminently the case at Moores Creek, where the flows are very heavy. Some of the smaller accessions were doubtless from the valley of Rattlesnake Creek, two irregular and somewhat extensive basalt dikes occurring about the forks 2 miles above the mouth. Just below the forks of Smith Creek, also, is another dike of the same composition as the foregoing, but with diabasic structure. Indeed, in the region crossed by trail from Boise to the Trinity Lakes and Rocky Bar, basalt and diabase dikes are numerous. The most isolated occurrence of diabase is as a complex near the head of Falls Creek, outcropping on the lower slopes of the Trinity Range. The outcrop is several hundred feet across, of irregular outline, and forms a conspicuous red and blue patch in the gray of the mountain side. The dikes, both of basalt and diabase, all cut gray granite.

A gray quartz-porphyry showing a groundmass of micropegmatite with porphyritic crystals of feldspar was observed on the Boise from a point 2 or 3 miles below the entrance of the South Fork, to Trail Creek. Its extent beyond this is undetermined.

On the ridge north of the stream flowing from the Trinity Lakes the granite is cut by a body of quartz-diorite porphyrite, of a finely crystalline matrix and porphyritic feldspar. Between here and Rocky Bar, a distance of 8 or 9 miles, there are several occurrences of this rock or the quartzless variety, diorite-porphyrite. Within the same region, also, are occasional dikes of basalt or augite-andesite, which was not determined. In the Red Warrior camp, 2 miles southwest of Rocky Bar, is also a light-gray to white, finely crystalline or microcrystalline quartz-porphyry, similar to the Leadville porphyry. Grains of pyrite occur disseminated through it. The specimen examined was from the dump of an inaccessible tunnel by the roadside, a mile or so northwest of the center of the camp. From the quantity, one may infer it to be a rock of importance, at least for this mine.

In the Jim Blaine mine, a mile and a half east of Rocky Bar, there occurs a dike of basalt carrying a few rounded quartz grains. Its relation to the ore body was not altogether clear.

On the eastern side of the high divide between Rocky Bar and Atlanta, on the lower half of the slope, there occurs in the granite a succession of enormous dikes, 100 to 200 feet wide, of quartz-porphyry and a quartz-bearing syenite-porphyry, which extends northwest several miles across country. What were probably similar occurrences were seen at a distance in several of the ranges about Atlanta, but the localities were not visited. The prevailing trend of the ledges is east and west, a direction often observed also in the granite and ore bodies of the Rocky Bar and Atlanta districts. The dikes dip 45° or 50° N. to vertical. Between them is the ordinary granite of the country, often appearing like included masses. The quartz-porphyry is typical, a felsitic groundmass, with small irregular bodies of quartz and occasional small crystals of feldspar, the general tint of the rock being a decided pink. In the quartz-syenite-porphyry, also pink in effect, the quartz is distributed sparingly through a finely crystalline groundmass, while the orthoclase is developed in prominent porphyritic crystals. The basic mineral is usually finely crystalline, and in aggregates mottles the rock a dark-green. The relation between the quartz-porphyry and the quartz-bearing syenite-porphyry was not determined, although they evidently occur in close association with each other.

In Montezuma Gulch, east of Atlanta, between a third and half the distance to the head, is a narrow dike of aphanitic syenite (lamprophyre) similar to that occurring in the Flannigan mines on the west slope of the Boise Range. Its lineal extent is unknown.

In the Monarch mine, on Atlanta Hill, is a narrow dike of probable decomposed syenite (lamprophyre), but the specimen collected was too much altered to permit satisfactory determination.

In the Tahoma mine is a dike of white, decomposed porphyry, also altered beyond identification. Its thickness varies from 25 to 50 feet on the several levels. In one or two places the vein has been slightly thrown by it. The general trend of the dike is N. 26° W.; the dip southwest. All the eruptives of the Atlanta district not infrequently contain pyrite in minute crystals; the influence of the dikes upon the mineralization of the veins, however, is undetermined.

In the granites of the Sawtooth Range, on both the east and west sides, are frequent, narrow, fine-grained dikes of aphanitic syenite (lamprophyre), and what is likely a nearly related rock with porphyritic crystals of triclinic feldspar. Quartz-porphyry was also observed on the east side of the range in the slopes west of Upper Redfish Creek. A large body of this rock appears in a side gulch entering the main valley about a mile above the lake, and, from their appearance at a distance, similar bodies are believed to exist at various points along the range front.

Rhyolite (aporhyolite) occurs as a large irregularly shaped dike in a hill on the east side of the Salmon Valley, 10 to 12 miles above the sharp bend of the river to the east and 4 to 5 miles above the Lower Redfish Lake. The rock is banded gray and brown, the brown often having a glassy or resinous appearance, the gray being very close textured, and marked with fine striations, as though of flow structure. The prevailing rock of the country is the usual gray granite.

In the morainal débris about the Lower Redfish Lake, derived from the Sawtooth Range, besides the more common pinkish granite, there are numerous bowlders of quartz-diorite-porphyrite—a dark-gray rock with porphyritic plagioclase and much sanidine.

A large body of quartz-porphyry of greenish-gray groundmass with feldspar in prominent crystals forms a high hill in the forks of Valley and Stanley creeks. The general trend of the mass is north and south. On the east side of Valley Creek, about 4 miles above the mouth of Stanley, is a prominent body of quartz-bearing syenite-porphyry, the groundmass of which is micropegmatite with a very small amount of quartz. The rock is greenish-yellow. The porphyritic constituent is orthoclase. Near the head of Valley Creek, by the roadside, are numerous small dikes of dark-gray, altered diabase, cutting granite.

The main Sawtooth Range was crossed only at one point, at the head of Upper Redfish Creek; but in the occurrence of eruptives in the débris from the portion north of this, and in the veinings that may be observed from a distance, there are evidences of a number of varieties, among which the quartz-diorite-porphyrite above described is an important type. The veinings mentioned may be seen in all parts of the main range, now as the chief mass of one of the rugged peaks, now in a broad band of color, running along the precipitous sides in contact with a pink-colored granite which here seems to constitute the mass of the range. North of Cape Horn the range has the appearance from a distance of returning to the gray granite as its chief constituent—the granite which occurred at the southern end where crossed and which is the prevailing rock of the great mass of mountains to the west. East of the Sawtooth the gray granite passes into the high range forming the divide between the main Salmon and its Middle Fork, and into the mountains east of the Salmon River and Valley Creek. In nearly every portion of its area, however, it is frequently cut by eruptives of one kind or another.

East of the trails to Sheep Mountain and Loon Creek, in the direction of Bonanza, great bodies of eruptives appear in the lofty peaks and intricate ranges there present. These were all beyond the route of reconnaissance. On the line of exploration, however, on the several heads of Beaver Creek, and in the gulch leading down to the mining camp of Sea Foam, are numerous dikes of diorite-porphyrite, crystalline-granular throughout, with porphyritic development of the feldspar. On Beaver Creek, also, is a white or light-gray, fine-grained variety of quartz-porphyry of frequent recurrence; magnesian mica is quite prominent throughout this rock. Other quartz-porphyries differing somewhat from the foregoing in the amount of quartz or the porphyritic development of their feldspar appear in dikes of greater or less extent in the mountains between Beaver, Bernard, and Loon creeks. On the latter stream they are particularly abundant. They all cut the granite of the region.

In the canyon of Loon Creek, 3 miles below Oro Grande, as dikes in a coarse, pink granite, occurs a finely crystalline quartz porphyry of a color closely resembling that of the granite. The orthoclase of the eruptive may be porphyritic, while the quartz usually occurs in minute grains in the groundmass. Both granite and eruptive are conspicuous features of the region's geology.

In the same vicinity occurs also a dark-gray quartz-diorite-porphyrite of a totally different facies from the pink. The crystallization of this is in two or three degrees of coarseness, white feldspar crystals being prominent, though not large. The rock is of extensive outcrop, and occurs in heavy masses, in places apparently resting upon the granite of the region; but such an appearance may be due to an inclined position of the dikes in the granite and to the relative position of their outcrops in the precipitous walls of the canyon. The rock again appears about 3 miles farther down the canyon, from which point it may be followed in continuous outcrop for a distance of 8 miles or more downstream—beyond Beaver Ranch—and to the east far toward the summit of the high divide between the waters of Loon and Camas creeks.

On the east side of this divide, along the West Fork of Camas Creek and on the western branches of the Middle Fork, is another dark-gray porphyrite with many feldspar and quartz crystals and containing numerous small fragments of dark-green rocks. A fluidal structure is clearly shown under the microscope, though in the hand specimen it would hardly be suspected. The rock forms the chief outcrop in the region mentioned, but a gray rhyolitic tuff is also present at intervals. Well down on the West Fork of Camas Creek is also a yellow to yellowish-red quartz-porphyry, outcropping in three or four heavy ledges 50 to 100 feet wide. This porphyry is in part solid, in part somewhat vesicular. Apparently it cuts the gray rhyolite, but their relations were indistinct.

On passing from the area of gray rhyolite and associated quartz-porphyry near its eastern edge, there appears at a point a mile and a half above the mouth of the West Fork of Camas Creek a second prominent series of lighter-colored rhyolites and rhyolitic tuffs in heavy beds striking about N. 15° E. and dipping 30° E. These beds are continued far into the mountains both north and south, and in a northern direction there appears at a distance a tendency in the strike to bend to the northwest, as though the series formed the eastern edge of a dome-shaped mass of eruptives. This, however, is indefinite. The relation of this second series to the dark-gray rhyolites and quartz-porphyries is unknown, their contact being covered at the point traversed, but it is quite possible that the tuff rests upon the gray rhyolites and is in turn overlain by the lighter rhyolites with which it is associated. The tuff is bedded, and consists of a soft, white, green, or purple matrix, with included angular fragments of rhyolite from a half inch to 2 feet in diameter. It is, at a rough estimate, 200 to 300 feet thick, and extends along the creek a distance of one-quarter mile.

Banded rhyolite of somewhat varying texture and solidity apparently overlies the tuff, occurring in nearly continuous outcrop for a half mile farther downstream. Other tuffs succeed, with their included rhyolite fragments, and a matrix rather less friable than that of the lower beds, forming the last outcrops along the West Fork. Across main Camas Creek from the foregoing series, however, and just below the Three Forks, is a very similar succession of outcrops, rhyolites and rhyolitic tuffs in alternating bands 500 to 600 feet thick, often repeated, striking about N. 60° E., with a dip of 10° to 20° SE. Although this series is similar to that on the West Fork and may be the same, direct continuity was not shown. This series is continued with variation to the head of the East Fork of Camas Creek (Silver Creek), and beyond into the ranges bordering Prairie Basin on the west. On Silver Creek several varieties of the rhyolite appear, together with other rocks more or less nearly related. Most of them occur in heavy ledges, walling in the canyon with ragged, precipitous sides. On the north side of Silver Creek, a N. 60° E. strike and northwest dip are maintained by the eruptives for a distance of 7 or 8 miles up the creek. On the south side, in the high ridge southeast of the Three Forks, the general dip is southward, with more or less gentle flexing. On the upper portion of the creek strike and dip vary somewhat.

Of the foregoing rocks, the tuff is of widest occurrence, and on both sides of the divide between Silver and Panther creeks it occurs in heavy beds, underlying the summit itself.

The Middle Fork of Camas Creek was not examined. From the Three Forks, however, there could be seen a few miles up, between the two chief branches, an enormous butte of eruptive rock, appearing like the remnant of a great lava-flow. It is of a very dark-brown color, and at a distance resembles basalt. At other points in this canyon the eruptives of the East and West forks seem to prevail.

In the gravel wash in the vicinity of the Three Forks were two varieties of rhyolite (aporhyolite) resembling that found in the upper valley of the Salmon River. These may have been brought down any of the streams.

From the foregoing it is evident that the region about the Three Forks of Camas Creek for a radius of 15 to 25 miles has at various intervals in the past been one of intense eruptive activity. The series of rhyolites and tuffs extends entirely across the drainage system, and from the high divide on its south to fully 10 miles below the junction of the three main creeks. Beyond this area, particularly to the west, rhyolites and other eruptives still occur over vast areas; to the east, while present, it is not to such complete exclusion of the primary rocks of the country.

The high mountains forming the divide between Prairie Basin and Yellow Jacket Creek, a tributary entering Camas Creek several miles north of the Three Forks, are largely composed of granite and of quartzitic and micaceous schists, having a prevailing strike between N. 30° E. and N. 20° W., and a dip generally away from the center of the range, though often very steep or even vertical. The slates, which alone were crossed on the route traveled, are frequently cut by eruptives, in some instances at angles with the strike and dip, in others coincident with them.

Among the eruptives are the following: At the entrance to Fourth of July Canyon, a massive body of pink, banded rhyolite, in appearance a sheet 50 to 60 feet thick, with a strike north and south and dip E. 20°, about that of the schists of the locality. The sheet extends for several miles along the eastern front of the range, with little apparent interruption. About 2 miles above the mouth of the canyon is a second dike of rhyolite, 20 feet wide, with finely developed dull-yellow feldspar crystals, one-half to 1 inch long. This is confined between slates, with about the same strike as they, but whether with coincident dip is uncertain. One mile above this is another eruptive of similar nature, striking N. 25° W., cutting through the slates. The summit of the divide carries a prominent outcrop of andesite, containing hornblende, mica, and augite. This rock is pink and coarse grained, with a white feldspar particularly conspicuous, abundant, and evenly distributed. The outcrop extends for a considerable distance along the range, and has somewhat the appearance of an irregularly shaped dike, or, in places, of a sheet lying upon the summit of the range. About a mile and a half down the western slope of the divide, on the trail, is a narrow dike of coarse, yellowish-gray quartz-porphyry with much orthoclase in crystals averaging an inch in length, many of them twinned on the Carlsbad type. A trachyte, finely crystalline and yellowish-gray, occurs just beside the porphyry and to the west of it, of apparently about the same width as the latter. A second dike of trachyte, similar to the first, about 15 feet wide, occurs 2 or 2½ miles lower down the valley, striking N. 15° W. and dipping W. 70°. The inclosing quartzites here dip 70° E., but it is not certain that the dike cuts them at all points. Still another dike, of dark-gray quartz-porphyry carrying hornblende and biotite with microspherulitic groundmass, occurs just above the confluence of the two forks of Yellow Jacket Creek. The foregoing eruptives all cut quartzite and mica-slates of the same kind as those on the east side of the range. The strike of the schists on the west side of the range varies from N. 15° to 75° W., with westerly or southerly dip. The eruptives seem generally to follow them in strike, but in dip they sometimes have the appearance of cutting across the stratification. On the main Yellow Jacket Creek, between three-quarters of a mile and 2 miles above Yellow Jacket camp, are numerous dikes of rhyolite similar to that at the entrance to Fourth of July Canyon. The dikes form enormous outcrops on the mountain sides, and seem to have a general trend N. 15° E. with a westerly dip of 50° or more. The dikes cut the quartzites of the country, and are locally so numerous that the slates appear as included bodies.

The eruptive rocks of the immediate Yellow Jacket mining district embrace the following: A quarter mile above the camp, syenite, a fine-grained, yellowish-gray rock, occurring apparently as a thin band (dike) between strata of the schists, which constitute the country rock, and with them striking N. 15° E. and dipping W. 45°. Dikes of quartz-porphyry occur between several of the veins in the Columbia ground. They are usually narrow, but of undetermined lineal extent. This rock is locally termed "porphyry" by the miners. A third rock, which occurs in numerous narrow dikes throughout the region, is a dark-gray mica-diorite. A fourth and oft-recurring eruptive is a syenitic variety rich in mica, known as minette; and a fifth, more hornblendic, is almost identical with the rocks mentioned above as aphanitic syenite. These rocks are typical of the complex group called lamprophyres by some petrographers. Their occurrence here is in various dikes, and they are regarded by the miners, whether correctly or not the writer can not say on personal observation, as indicative of an ore body near at hand. In local parlance they are termed "syenite." Usually, so far as opened, when occurring next to the veins they form their foot walls, while the quartzite of the country or one of the other eruptives usually forms the hanging walls. A sixth eruptive, capping the high ridge northwest of the Columbia mines, is a very coarse quartz-porphyry, of the same general character as that a mile and a half west of the Panther-Yellow Jacket divide. A seventh rock is an altered diabase, observed as a dike in the Yellow Jacket mine. The general trend of the dikes is with the strike of the slates and main veins, N. 60° E., with perhaps occasional local variations.

On Panther Creek, about a mile and a half below Fourth of July Canyon, is a small deposit of a soft, white tuff containing angular fragments of rhyolite and possibly other eruptives. Its derivation is unknown. Another and more extended area of the same nature occurs on Napias Creek a little below California Bar, where it forms the hills both north and south of the stream for a mile or more back. On Napias Creek, about 2 miles above its confluence with Panther or Big Creek, are coarse-grained syenites of two somewhat different types. They occur in apparently heavy bodies of undetermined outline in the bird's-eye granite of the region.

A half mile above California Bar, on the northern side of the creek, is a small dike of either pyroxene-andesite or basalt, green and pink in color, the latter probably representing a stage of decomposition. Other eruptives occur in heavy masses in the vicinity of California Bar, but time did not permit examination.

From the Salmon City Valley to the Pahsimeri, the canyon of the Salmon is for long intervals cut through eruptives closely resembling in composition and occurrence the series of rhyolites in the Three Forks region. At the entrance to the canyon, about 6 miles above Salmon City, are beds of fine rhyolitic tuff, yellow, green, and pink, dipping 10° or so north, downstream. Underlying, and forming the actual entrance to the canyon, are 500 to 1,000 feet of pink and gray, banded rhyolite, somewhat spherulitic, and of varying hardness. The dip is 15° to 20°, its direction varying between northeast on the east side of the river and northwest on the west side. In fact, from their position, the rocks of this series appear to belong to a volcanic dome the center of which is at some point considerably to the south.

Beneath the rhyolite, with same dip, is an equally thick, very dense, dark gray, banded rock, probably augite-andesite. Its exterior is usually green or pink, from the weathering of some contained mineral, possibly of chloritic nature. About 3 miles farther up the canyon the andesite is followed by a second rhyolite, which appears from beneath the former in a great half dome, dipping north. The rock is light gray, with a vitreous groundmass, and, like the other rhyolite, displays marked banding. The foregoing rocks constitute the mountain masses on both sides of the canyon to the highest elevations, the gorge cut through them being sharp and rugged.

Beneath the rhyolite last described is a heavy series of dark-gray quartzites (Algonkian?), which continue for 8 or 9 miles along the canyon to a point about 18 miles above Salmon City. Here an augite-hornblende-andesite abruptly cuts the schists, the rock varying from dark-gray to pink, with a microcrystalline groundmass and porphyritic feldspar. In this same vicinity, judging from fragments, is also a quartz-bearing porphyry, possibly related to rhyolite. The andesite extends along the east side of the river for at least a half mile, when it passes beneath quartzite débris from the mountains above. On the west side of the river, a mile above its lower edge, it is overlain for a short distance by 60 to 75 feet of lava, showing columnar structure. Between 21 and 22 miles above Salmon City, at the upper end of Edson's ranch, occurs a basic, often amygdaloidal basalt, almost free from feldspar, and carrying a green or red alteration mineral. Toward the lower end the outcrop pitches gently south, but changes to an equal amount to the north, or becomes horizontal, a short distance on. There are two varieties of the rock, one hard and dark-gray, the other pink, light, and vesicular. It appears more as a valley rock than one entering into the structure of the higher parts of the mountains proper.

About 28 miles above Salmon City, schists again appear on the east side of the river for 2 or $2\frac{1}{2}$ miles, to be then again cut out by the basalt, which on the west side of the river has continued all along. At a point 31 miles above Salmon City another series of rhyolitic and closely related rocks appears from beneath the basalt, with a general pitch of 20° to 30° northward, or downstream. The eruptive first beneath the basalt is a banded, pink trachyte. This is apparently 300 to 400 feet thick. Beneath it, with conformable pitch, is a pink rhyolite, immediately succeeded by a brilliant-green volcanic breccia. The matrix of this last rock consists of coarsely crystalline or microcrystalline rhyolitic material; the fragmental components are either green or pink rhyolite, or possibly trachyte, an associated rock in this region. The fragments are small to large, angular or slightly rounded by attrition, and the pink variety of rhyolite or trachyte is occasionally found vesicular. There is, altogether, a close resemblance between the fragmental constituents of this breccia and the eruptives occurring in association with it. Stratification is occasionally strongly developed in this breccia, notably at a point about 38 miles above Salmon City. The dip is to the northward, the strike bending between northeast and northwest.

The series of eruptives thus enumerated closely resembles, in field appearance at least, that observed on Silver Creek, 20 or 30 miles west. The succession of the various types may differ somewhat in the two localities, but similar composition, structure, and general occurrence are clearly perceptible.

The bright-green, rhyolitic breccia extends along the river, with but few interruptions, to the Pahsimeri Valley, a distance of nearly 12 miles, forming the canyon walls to a height of 500 or 600 feet or receding to more distant hills according to the extent of the erosion. It is occasionally cut by one of the other eruptives, or its continuity may be interrupted by an anticline of Cambrian (?) quartzite, common in the region between here and Challis. Three or 4 miles below the mouth of the Pahsimeri the breccias are capped by heavy lava-flows, probably of hornblende-andesite, a rock of frequent and conspicuous occurrence between here and the upper end of the Pahsimeri-Salmon Valley, 3 miles above the mouth of the former stream. In the latter locality the andesite abuts against a quartzite anticline, and also appears on the slopes of hills to the east at much higher elevations. West of the Salmon an old valley in the northern slope of the quartzite ridge is filled with eruptives, appearing at a distance to be partly breccia and partly andesite. These eruptives again appear south of this anticline, still abutting against the quartzites or capping the lower spurs of the ridge. From here they continue to the Challis Valley, forming river bluffs and adjacent hills. They surround the quartzite anticlines of this region, and, locally, rise slightly upon their sides. At the lower end of the Challis Valley, at the entrance to the canyon of the Salmon, the east bluff of the river shows several hundred feet of interbedded breccia and andesite, the layers often presenting uneven lines of separation, or wedging out entirely, all without semblance of regularity. Just back of the town of Challis is the last exposure of the breccia observed along the route.

It was impossible to delineate the area occupied by the breccias; their extent along the Salmon has been given—35 or 40 miles. East of the river they seem to be limited within a few miles (2 or 3) by the quartzites or other rocks of the Salmon Mountains, excepting, perhaps, along the Pahsimeri River, up which they extend for at least 4 or 5 miles, possibly farther. West of the Salmon the extent of the breccias, for the southern portion of their area at least, is considerably greater, possibly 5 or 10 miles from the river in some instances. The nature and configuration of the deposit as a whole are difficult to understand without detailed work. Rhyolites, andesites, basalts, and breccias are associated at one point or another, and, as in the region lower down the Salmon and, again, about the Three Forks of Camas Creek, would seem to form yet another great volcanic center.

At the upper end of the Challis Valley, along the foot of the range inclosing it on the south, occur enormous bodies of a gray basic andesite containing both hornblende and augite and grading into basaltic varieties. Their outcrop on the river is several hundred feet in height and between 2 and 3 miles in length, and they apparently extend for a considerable distance into the hills on either side.

About 6 miles east of the Salmon River are several small exposures of greenish-gray andesitic or rhyolitic tuff dipping about 15° northeast.

The divide between Warm Spring and Antelope creeks is capped with a thin layer of bright-red rhyolite. The groundmass is homogeneous, and the porphyritic mineral, chiefly feldspar, at a minimum. The rhyolite overlies green and yellow tuffs, similar in appearance to those seen elsewhere in the Challis Valley. The entire series seems to have a gentle rise from periphery to center, forming a dome 2 or 3 miles across.

The Challis Valley abounds in eruptives of one kind or another that may be seen in outcrops, low domes, buttes, ridges, etc., scattered over the greater part of its area. It is preeminently an area of volcanic rocks modified by erosion and the deposit of materials derived from the inclosing ranges, sedimentary and crystalline.

In the Wood River Valley, both in the vicinity of Ketchum and for several miles above and below, are numerous dikes or other irregular masses of andesite, embracing the hornblende-mica and hornblende varieties, with feldspar porphyritically developed. The larger bodies are several miles in extent, showing in massive outcrops in the valley above the town, and below, extending well into the ranges west of the river. On Deer Creek, 8 miles below Ketchum, there is a small outcrop of andesite, but its continuity with bodies higher up the main stream was not proved. The general trend of the larger bodies of eruptives is north-northeast; the pitch, westward. The eruptives cut the great shale and limestone series forming the mass of the mountains in this district. They seem, however, to be quite independent of the ore bodies of the Wood River district, and therefore to have had little if any influence upon their mineralization; but the entire region is one of marked metamorphism and mineral impregnation.

A few miles south of the Wood River district lie the plains of the Snake River and its tributaries in this region. Among the latter, the valley of the Big Camas is underlain by heavy sheets of augite-andesite, which also extend to the higher lands along the base of the mountains both north and south. A prominent sheet appears along Camp Creek, extending from 1 or 2 miles below Doniphan to the mouth, resting upon granite, which is exposed at one or two points. The flow has a pitch to the south, as though acquired by the general inclination of the underlying surface. Bodies of this eruptive also appear to the northeast of Doniphan. The ridge south of Camas Prairie, separating it from the Snake Plains, appears from a distance to be largely composed of augite-andesite, granite, by report, being the other constituent. On High Prairie, at the head of Big Camas Creek, are still other heavy outcrops of augite-andesite; and indeed it is quite likely that the entire floor between the granites on the north and south is underlain with it, though now covered with wash from the mountains.

These andesites are probably directly related in occurrence and composition to the basalts of the Snake Plains and of the Boise country a short distance further west, and to those within the mountains to the north.

In the granite of the foothills between High Prairie and the Boise are several dikes of fine and coarsely crystalline quartz-porphyries. The dikes are narrow and have a lineal extent of not over 1 or 2 miles. Similar dikes have already been cited on the western slopes of the Boise Range and along the Boise Canyon within the range itself.

From the road from Nampa to Silver City, across the broad plains of the Snake, may be seen numerous exposures of basalt in sheets horizontal, or buckled in low dome-like elevations from a half mile to 2 or 3 miles in diameter and 100 to 500 feet high. Of the latter occurrence Initial Point is an illustration. In some instances erosion has begun to take effect at the center of the dome, probably from a greater abundance of cracks than elsewhere. Along the Snake, in the vicinity of this road, the lava shows in precipitous walls, resting upon the Tertiary beds which are exposed in the lower portion of the bluffs. South of the river, in both valley and mountains, are enormous bodies of eruptives of various types, in dikes, sheets, and irregular masses. In one or two localities certain of them carry opals of gem value.

The eruptive rocks of the De Lamar and Silver City district embrace rhyolite, diabase, and basalt. The country rock of the De Lamar mine is rhyolite of rather fine, even texture. The general color in the mines and in many places outside is white, but a darker, greenish variety also occurs. The extent of the rhyolite body is said to be about 5 by 3 miles, the greater dimension lying northwest-southeast. In the Black Jack mine the country rock is also rhyolite, very similar to that of the De Lamar mine, but whether of the same body was not ascertained. The rhyolite becomes highly kaolinized in a certain part of the mine, and is then mistaken by the miners for granite, and is so called. Diabase, called by the miners "trachyte," also occurs as a dike in the Black Jack mines, and basalt appears as a surface flow a half mile below the mine on the mountain side. The occurrence of these eruptives and their relations to one another are somewhat obscure, except that the diabase and basalt are later than the rhyolite.

The effect of the eruptives upon the mineralization of the ore bodies will require detailed observations in field and chemical laboratory. The rhyolite near the veins nearly always shows some degree of mineralization.

GENERAL STRUCTURAL FEATURES.

The mountains of the southern half of Idaho are developed en masse or in ranges the lines of which have a north-northwest or an east-northeast direction. The development en masse is confined to the western half of the State; the system of ranges, although not confined to the eastern, is nevertheless its chief order of structure. In the mountains en masse it is quite probable that the elevations were attained in pre-Paleozoic times, while the range system may have undergone several periods of development, to late Tertiary.

The range of primary importance is the Continental Divide. This pursues a somewhat irregular trend along the State's border, but east of the Lemhi and Salmon valleys for a distance of nearly 100 miles it has a general course between N. 30° and 50° W., which is also that of its structural axis. The quartzites and schists showing in its western face are locally crumpled, but the prevailing dip is 50° or 60° SW. Regarding the nature of the uplift, it is believed to be an anticline, but time permitted an examination at only one point on the western side.

The development of the Continental Divide, and also of the ranges to the west, has doubtless continued in varying degree from early times to the Miocene, for nonconformities occur within the crystalline series, and in the Tertiary beds of the Salmon City Valley folding of considerable importance may be seen, along both the Lemhi and Salmon rivers, and east of them, in the region of Kirtley and Carmen creeks.

Parallel with the fold of the Continental Divide in Idaho are numerous others, represented in the Lost and Salmon River ranges and those immediately to the west. The greater part of these ranges lay so far beyond the route of survey that their structure could be seen only remotely. Their northwestern ends, however, were in many instances observed along the Salmon and Lost rivers in the passage from Salmon City to Wood River.

The general strike here encountered was about N. 35° W., the dip now northeast, now southwest. This is shown in the areas of Algonkian (?) schists and quartzites between Salmon City and the Pahsimeri, and in the several anticlines that occur in what have been provisionally regarded as Cambrian quartzites, extending along the Salmon River from a point 10 or 12 miles below the Pahsimeri to Challis. Many of the anticlines, having been cut by the river near one end or the other, show both exterior and interior structure, presenting canyon walls sometimes in sheer precipices 1,000 to 1,500 feet high. While the majority of the folds passed are anticlines, others are quite possibly monoclines, the result of faulting, though this was suggested rather than proved by minor strike faults and by the general configuration of some of the ridges observed.

In the mountains of the Wood River region the development is again upon a N. 30° W. trend; the structure is anticlinal, the axis of elevation passing east of the river, in the mountains between this and the Lost and Salmon rivers. The northern extent of this anticline is unknown; to the south, as observed from a distance, it apparently sinks beneath the Snake Plains. Between the Wood River ranges and the main Sawtooth, about the head of Salmon River, is a complexity of folds, the details of which could not be deciphered from a distance.

In the region thus far described the folds which have governed the range structure are particularly clear by reason of the presence of the sedimentaries, altered or unaltered. In the region of granites to the west, however, structural lines are more indefinite, but northwest-southeast trends, with dips to southwest or northeast, may still be found in sufficient number to warrant the belief that certain of the ranges, at least, were developed on the same lines as those, more pronounced, in the eastern portion of the State. Among these are the Sawtooth and Boise ranges, although the former has been greatly influenced by the texture of the rocks entering into its composition. Still, even the lines due to this influence were probably governed originally by the direction assumed in the fissuring, which controlled the trend of the rocks of eruptive origin. This is particularly noticeable along the crest of the Sawtooth.

An east-northeast structure is suggested at many points in the granites of the western half of the State. It is often but incipient, shown chiefly in lines of jointing, in occasional strikes of foliation planes, or in the trend of fissures, but the constant recurrence of these features indicates its prevalence, and that the forces which produced it acted with greater or less energy over a large portion of the western half of southern Idaho. Many local instances of the lighter effects occur about Rocky Bar and Atlanta, but the most pronounced illustration is the transverse ridge which forms the divide between the main Salmon and its Middle Fork. This ridge is an east-northeast anticline, the strata, whether granite or schist, quartzites or limestones, all showing the dip away from the center. The structure is traceable, with some obscure intervals, to the drainage of Camas Creek, within a few miles of the Yellow Jacket mining district. The interruptions are due both to complex folding and to the presence of large bodies of eruptives.

The divide between the waters of the Salmon and Snake rivers can not properly be considered a range in a strictly structural sense, for it is evidently composite in character. For 40 or 50 miles east of the Sawtooth it is probably a complex of folds. East of this it is formed by the transverse links between the several ranges of northwest trend, or by the ranges themselves, and to this combination is due the extraordinarily irregular trend of one of the most important watersheds of the State.

West of the Sawtooth Range, with one or two exceptions, the loci of the divides between the drainage basins are doubtless partly due to early accidents of structure—folding, faulting, jointing—but in the wide distribution of the homogeneous granite the structural features are only slightly pronounced and erosion has exerted its full influence in modeling the configuration of to-day. The Boise Range, however, is clearly developed on northwest-southeast lines of structure, a part of the general system of the State.

The intermontane valley at Horse Shoe Bend, Payette River, like that at Salmon City, shows several flexures in its Tertiary beds. The most important is an anticline running diagonally across the center of the valley southeast to northwest, the other folds being subordinate to this, but developed in the same lines.

The ridge between the valley of Big Camas Creek and the Snake, which has a nearly east and west trend for 50 or 60 miles, shows from a distance a number of broad, gentle flexures, but their axial lines were not clearly visible from the route of travel. The western half of the ridge, at the head of Little Camas Creek, is a part of the general system of elevations which belongs to the mountain region just north.

The Owyhee Range in southwestern Idaho belongs to the type of Desert Ranges. It is of granite, cut by eruptives, with Pleistocene and probably Tertiary sediments occupying the small interior valley of Reynolds Creek. The sedimentary beds show slight folding, indicating movement in later times, but the range received its greatest development prior to the Tertiary.

MINING DISTRICTS.

GOLD AND SILVER.

The mining districts of the precious metals visited in the reconnaissance include the region about Rocky Bar and the adjacent Red Warrior camp, together known as the Bear Creek district; the celebrated Atlanta lode and its attendant veins; the Yellow Jacket camp; the California Bar, Leesburg, and Lemhi placers; Wood River; and the Silver City and De Lamar camps in the Owyhee Range. All are gold or silver producers in varied degrees, and have histories dating back from five to thirty years, with yields reported in some instances as enormous. At the present time the industry throughout the State is, with a few conspicuous exceptions, at a low ebb.

THE BEAR CREEK DISTRICT.

Under this are included the Rocky Bar and Red Warrior camps and the region immediately about—in all, an area 3 or 4 miles in diameter. Placers formerly extended along the main stream to points considerably beyond, but are now worked by a few Chinese only. The town of Rocky Bar lies well within the southern limit of the great mountain mass of Idaho, in a deep, narrow gulch near the junction of Beaver, Steel, Feather, and Bear creeks, from the latter of which the district receives its name. The elevation at Rocky Bar is 4,800 feet; the surrounding ridges rise to 6,500 feet; the crest of the divide between this and the Atlanta district, to 8,500 feet; while Mount Steel, the highest peak of the region, attains an altitude of 9,500 feet. Communication is by a good wagon road to Mountain Home Station, on the

Oregon Short Line, 64 miles distant. Red Warrior Camp lies from 1 to 2 miles southwest of Rocky Bar, about the heads of Red Warrior Creek, a short tributary of Bear Creek. The dividing ridge between

FIG. 39.—Sketch map of Bear Creek district.

the two camps is 500 to 1,000 feet higher than Bear Creek, and has a general trend northwest. A wagon road leads from the camp to the main thoroughfare from Rocky Bar, following the stream.

The country rock of the district is the gray granite entering so largely into the constitution of the Boise and Sawtooth ranges. It is an aggregate of feldspar, quartz, and mica, chiefly biotite. The feldspar occasionally becomes porphyritic. The granite is inclined to the massive form, but it is not altogether devoid of foliation, which in some instances strikingly suggests bedding. The prevailing strike of the foliation planes is N. 50° to 70° E., while their dip is generally toward the northwest. Minor flexures, however, occur and the entire country displays more or less fracturing, especially along the gulches.

Eruptives occur as narrow dikes in the granites. They include basalt or olivine-diabase, quartz-porphyry similar to the Leadville porphyry, and diorite-porphyrite. These eruptives were found at many points along the route of survey in localities not noted for the presence of ore bodies, and their occurrence in a region of metalliferous veins, notwithstanding they contain locally a greater or less amount of finely disseminated sulphide of iron, may have been quite without influence upon the mineralization of the district; indeed, their mineral contents may have been derived from the same source as those of the ore bodies. The quartz-porphyry of Leadville type which occurs in a tunnel in Red Warrior Gulch a short distance above the camp, and perhaps at other points as well, is particularly rich in pyrite.

The mineral veins throughout the Bear Creek district are of quartz, carrying auriferous sulphides below water-level and their alteration products and free gold above. Most of the veins strike and dip with the foliation planes; occasionally, however, they may be found crossing these. In width they occur up to 12 feet, or in some places there may be a number of small veins, separated from one another by narrow bands of granite, the whole forming a single mineral-bearing zone 5 to 10 or 15 feet wide.

The included granite, and also that for 4 or 5 feet on either side whether of zone or single vein, usually shows a marked change in appearance. The micaceous compounds, particularly the biotite, have suffered greater or less decomposition, with dissemination of the iron throughout the rock mass, and the feldspars have become kaolinized in considerable degree. The granite appears, in fact, as a crystalline aggregate of feldspar and quartz, more or less ferruginous and disintegrated. This is commonly designated "vein matter." Such a zone of alteration is visible not only in open cuts and mines but frequently with great distinctness along the outcrop of a vein.

The quartz is coarse to finely crystalline, occasionally banded, and carries the mineral either disseminated, or in bunches, or following the planes of structure in thin seams. The material varies in richness of impregnation and the sulphides themselves in gold contents. The prevailing sulphide is pyrite, but those of antimony, zinc, galena, and copper are also present. The last is rare, and the zinc and galena sulphides are of minor importance.

The gold of the district is light colored and is said to be worth about $14 per ounce, the ore running from $6 to $30 per ton without concentration.

The openings in the Red Warrior camp, with the exception of the Wide West, which has been closed during the past season, are hardly more than prospects, developed by tunnels. The veins usually follow the planes of foliation, the prevailing strike being between northeast and east, varying slightly. At one or two points a crushed condition of the granite was observed, with what appeared to be numerous small and irregular gash veins. A suspected dip of the fractured zone was 80° NE.

The mines about Rocky Bar embrace some of historic interest—the Old Alturas, Idaho, and Vishnu—while the placers in early days were almost phenomenal in their product. At present the industry is quiet, and although a depth of 700 feet has been attained in some of the mines, only the upper levels could be examined. Quartz veins occur singly and in zones; in the former case, of widths up to 5 or 6 feet, locally even greater; in the latter, including intervening granite, up to 20 feet, the entire zone being mined. The latter condition prevailed, it is said, in the Old Alturas mine. The trends of the veins in this camp vary, but one between northeast and southeast prevails; the dip is usually south, though in one or two instances north. The upper portions of the veins are sometimes broken over in one direction or the other, dipping as low as 30°; in the lower levels the dip is said to be nearer vertical.

THE ATLANTA DISTRICT.

This district lies 18 miles northeast of Rocky Bar, across the divide between the South and Middle forks of the Boise River, at the junction of the latter stream with the Yuba. The center of mining interest is Atlanta Hill, an eminence of 1,500 or 2,000 feet in the forks of the streams, the end of a spur from the Sawtooth Range to the east. The town of Atlanta lies in the valley of the Middle Fork, at the northern base of the hill, 5,500 feet above sea-level. From this side three gulches enter the valley, Montezuma to the east, Quartz in the center, and a short, nameless one to the west. The first two head quite at the crest of the hill. The southern face of the hill is not so deeply eroded, and at its base Grouse Creek flows in nearly direct line to the Yuba. Confined to the hill, so far as at present known, is the celebrated Atlanta lode and its associated fissures. At the east both hill and lode seem to end abruptly against a higher portion of the spur, while to the west, beyond the Yuba, the presence of the lode has never been definitely established. South of Atlanta Hill, beyond Grouse Creek, the mountains, still spurs of the Sawtooth, rise to heights of 9,000 feet, forming the divide between the Yuba and South Fork of the Boise. North of Atlanta, across the Middle Fork, uniting Greylock (9,000 feet) and another lower peak to the west, is a prominent ridge capped with morainal or other drift, the crest fully 2,000 feet above the present level of the river. The Atlanta Valley is a gravel flat 3 or 4 miles long by 1 or 2 wide. The mountains about are sharp and rugged, the canyons often impassable, with evidence of glacial action in nearly all their upper portions. The district is 50 or 60 miles within the southern limits of the great mountain mass of Idaho, and but for a good wagon road to Rocky Bar and thence to Mountain Home Station on the Oregon Short Line, would be difficult of access. Trails lead from Atlanta to mining camps about, among them Sawtooth City and Vienna, in the upper Salmon Valley. The latter places are connected by wagon road with the Wood River region, about 40 miles distant.

The country rock of the Atlanta district is gray granite. It is similar in composition to that of the Bear Creek district, but structural planes or planes of foliation are not so prominent as there. To the west and east, however, gneissoid granite does occur, and on the eastern side of the Sawtooth Range are some very fine examples of this type of rock. The rock of the Atlanta district affords many evidences of extensive fracturing, and it may be that to this is due the difficulty of recognizing planes of foliation.

The eruptive rocks of the region occur in dikes of greater or less prominence. In the range west of the Yuba and Boise rivers, and also in the high point in the elbow of the latter stream, are a number of dikes 100 to 200 feet wide, of fine to coarsely crystalline, pink quartz-porphyry and quartz-bearing syenite-porphyry. These extend several miles across country, their general trend being a little north of west, their dip 45° or 50° NNE. to vertical. Aphanitic syenite (lamprophyre) also occurs in a small east-and-west dike in Montezuma Gulch, halfway up, and again in the Monarch mine in Quartz Gulch. Other dikes of the same or different nature are reported cutting the Atlanta lode in the various mines, but they were all inaccessible. In the Tahoma mine, which is on one of the lateral fissures, a dike of white decomposed porphyry, 25 to 50 feet thick, cuts the vein at a very acute angle. Its trend is N. 26° W.; its dip, W. 45° to perpendicular. On some of the levels it has thrown the vein, but this is again brought into line within 50 or 60 feet by a second fault, approximately parallel with the dike, marked by a clay course.

The influence of the several dikes of the district upon the mineralization of the veins has never been ascertained by those who have had charge of the mines, and the time allotted to the reconnaissance, together with the inaccessibility of many of the mines, did not permit the examination requisite for the solution of such a problem. The dikes are evidently outflows of several periods, and assumed many directions of trend.

The central feature in the Atlanta district from the mining standpoint is a main lode having what appear to be numerous nearly parallel branches, spurs, or feeders, forming acute angles with it; at the ends

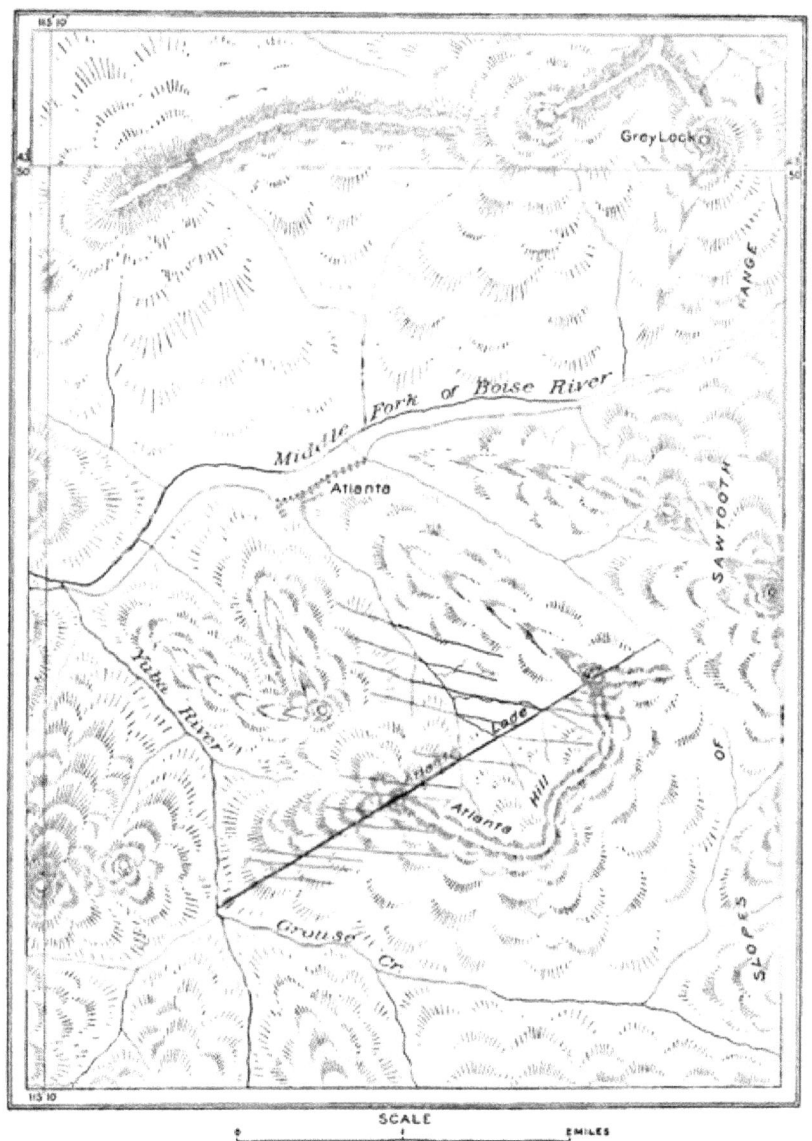

SKETCH MAP OF ATLANTA DISTRICT.

the lode itself may split. The main lode is known as the Atlanta; the spurs are designated by the names of the several mines located upon them. The outcrop of the Atlanta lode has an almost uniform trend of N. 50° to 60° E., and occupies a position on or near the summit of Atlanta Hill, running nearly with the ridge. Its eastern end is supposed to be near Montezuma Gap; it follows thence westward along the crest of the ridge to the drainage of Quartz Gulch, which it crosses about one-fourth to one-third mile below the summit, and again taking the ridge, continues on the crest to the western slope, down which it passes to Yuba River. The total length of the lode exposed and prospected is $2\frac{1}{2}$ miles; its continuation beyond this has not been proved. The width of the lode is said by those conversant with its early exploitation to vary between 50 and 150 feet, 75 being regarded a fair average for the whole. This width was the maximum observed by the writer, but only a very small portion was accessible to him. The lode is at present divided between four or five companies: one, of which Gen. W. H. Pettit is the manager, at the eastern end, owning 1,500 or more feet; a second, controlling a few feet only, coming next; the third, the Monarch, extending to Quartz Gulch; a fourth, the Buffalo, to the west of this; and the fifth, now the Atlanta Consolidated Gold and Silver Mining Company, occupying several thousand feet at the western end. The dip of the Atlanta lode varies in direction. At the eastern end it is to the N. 70° or 80°, becoming vertical at the western limit of the Pettit mine; in the Monarch mine a change to the south takes place; in the Buffalo the dip continues S. 45° to 70°; while for the western fourth of the lode, the position of the vein is again vertical or with but a slight southerly underlay.

The vein matter of the Atlanta lode is a more or less fragmental, clear to bluish-white quartz, and an altered granitic rock in which quartz predominates and the feldspars are highly kaolinized and the micas have generally disappeared. This second rock is doubtless related to the country granite, and has undergone the changes so common in rock adjacent to or in mineral veins. More or less clay is present locally, and some calc-spar in lenticular bodies. The entire mass of the vein is very friable and easily mined. The quartz is the chief metal-bearing constituent, and occurs from a thin seam to one 6 feet across, in one or several courses, on either wall or in the interior. Although a considerable proportion of the vein may contain ore in pay amount, it is said to be rare, if ever, that the entire width is mineralized to this extent. Again, it is often the case that narrower streaks contain as large amounts of gold and silver as the broader. The granitic ledge matter shows mineralization, but is usually regarded as too low in content to pay for mining. The calc spar is considered by some miners as a good indication of proximity to a rich ore body.

The ore, according to General Pettit, who has been familiar with the lode from its earliest exploitation, lies in shoots, four or five in number, along the length of the vein; one in the mine at the eastern end of the lode; one in the Monarch, pitching westward, toward the dike occurring near its shaft; one just beyond this dike, also pitching toward it, or eastward (this in the Buffalo mine); and one or two in that portion of the ledge now being opened by the Atlanta Consolidated Gold and Silver Mining Company. Moreover, in these shoots, the ore is said to be in the form of larger or smaller lenticular masses, which overlap one another in depth, occupying different positions in the width of the vein.

The Atlanta lode has the appearance of an original fissure the filling of which has been crushed by reason of movements belonging to a second period of disturbance. Mineralization may have been effected either at the time the original fissure was filled or subsequent to the fracturing that resulted from the second movement. The walls of this lode are horizontally, diagonally, or vertically slickensided; there is often present a strong clay selvage—almost always a trace of it; and the vein matter has been rendered friable by a marked degree of fracturing, the sharp, angular fragments often forming a brecciated mass, held together in a clay of fine siliceous cement. Slickensides and clays are also present in the interior of the vein.

The mineral contents of the vein are free gold and an auriferous sulphide of iron—pyrite or the related arsenical and antimonial compounds. Prof. J. E. Clayton,[1] from an early and more complete examination of the lode—chiefly, however, in the Monarch and Buffalo mines—reports the metallic contents as "gold, native silver, ruby silver, brittle silver ore, and sulphide of silver and pyrite. The brittle silver, or black antimonial silver, is the most abundant ore. Next in quantity and value is the ruby silver. The native silver and silver glance are found only in small quantities. The free gold constitutes from 20 to 40 per cent of the total value. The other minerals are iron pyrites in moderate quantities disseminated through the granular, friable quartz and the granitic inclosures of the lode. I saw but few traces of copper, zinc, or lead." Professor Clayton also adds, as the result of his observations, that "much of the quartz in this lode is comparatively barren. The rich streak of black sulphuret and ruby ore [probably in the Monarch and Buffalo mines] varies in width from 1 foot to 6 or 7 feet, and alongside of it is a zone of pay rock, equally as wide, that carries a good percentage of free gold with silver ore disseminated through it, making the pay streak from 2 to 15 feet wide and extending in length underground, in the Monarch and Buffalo claims, nearly 2,000 feet on the course of the lode." The character of the ore is said, however, to show material changes in the length of the lode.

As worked at the present day the average yield of the lode is about $20 per ton in gold. From this it varies in either direction. Ten

---

[1] Trans. Am. Inst. Min. Eng., Vol. V, p. 471, 1876-77.

dollars is regarded as the limit of profitable working, while bunches of ore of a value of several hundred dollars per ton are occasionally met with. The ore is now hand-sorted, a portion being milled in the camp, a portion concentrated and shipped out. In the early days, however, no effort was made to save ore, even the richest, that would not mill free gold; many of the dumps at the present time are therefore correspondingly profitable, and every season a portion of them is reworked by hand jigging or some other simple process.

The average proportion, in values, of gold to silver in the ores of the main lode is as 1 to 2; in the lateral veins north of the main lode, also as 1 to 2; but in some at least of those south the proportion is said to be as 3 to 1.

The fissures in Atlanta Hill transverse to the main lode have an E.W. to N. 65° W. trend, and dip north, south, or vertical, but always steep. Their structural relations with the lode are undetermined; some of them have the appearance of spurs, offshoots, or "feeders," while others may prove to be independent veins. They lie both to the north and south of the main lode, and those on opposite sides are in some instances so nearly continuous in their course that they may readily be suspected of being one and the same fissure. In no instance, however, has the passage of a lateral vein into the main lode been traced, except perhaps in the Pomeroy-Last Chance. Here General Pettit states that, with other objects in view, he traced the lateral directly into the main fissure, without a separating wall of any kind, the ground in the acute angle of the two veins being highly fractured.

The lateral veins rarely attain a thickness greater than 8 or 10 feet, and often but 1½ to 2 feet, or even less. Some of them are very rich in metallic contents, and carry mines of much importance. Among these are the Last Chance, Big Lode, and Tahoma, the first being to the south of the main lode, the others to its north. The ore of the lateral veins in a general way is similar to that of the main lode, consisting of free gold, native silver, and the sulphides already mentioned, in a quartz vein-stuff. Not all of the spurs show a fragmental, breccia-like character in the filling to the extent seen in the mines of the Atlanta vein, but the feature is not absent. The value of the ore of the transverse veins is said to vary considerably in the different mines, certain of them presenting material of low grade but of steady occurrence; others, material of high grade in smaller and more intermittent bodies. The relative proportions between the metallic contents also vary in the different veins, not only on the same side of the main fissure, but on opposite sides.

The Atlanta lode and its branches have yielded in past years enormous sums, and a favorable inference may well be drawn from history and observation as to the possibilities in the future. The lode is well located with regard to economic development, while timber and mining supplies may be obtained with comparative ease.

THE SHEEP MOUNTAIN DISTRICT.

This district is in the heart of one of the highest and most rugged mountain regions of Idaho. It is accessible only by trail, though a good mountain road is possible via Beaver and Bernard creeks from a point on the State wagon-road in the vicinity of Cape Horn. The center of the district is a hill 7 or 8 miles down Bernard Creek, north of the divide between the upper Salmon and the Middle Fork. The periphery of the district, however, includes the divide, and along this there has been considerable prospecting; owing to misinformation, it was the only locality visited, and the following description refers entirely to it.

The country rock is gray granite and quartzitic and micaceous schists, which in some layers are strongly calcareous. The schists occur both in series by themselves and as included belts, 1 to 100 feet or more wide, in the granite. The latter is more commonly the occurrence on the summit of the divide, while the schists in series are found chiefly on the northern slope. The bands of especial interest here are those in the granite. They may be traced for a mile or more, or may disappear after a few hundred feet, seeming to occur as irregular, isolated bodies. This appearance may be due to original differentiation in the material from which the granites and schists were derived or to included masses through faulting, the former perhaps being most plausible. The strike of schists, and of the granites where they show foliation planes, varies to all points of the compass, N. 15° to 35° W., N. 60° E., and east and west, being most common. This variation is due to local flexures, evidences of which are constantly recurring. The dip of the series in the divide is to the west or south, usually at a high angle.

Diorite-porphyrites and quartz-porphyrites occur in dikes throughout a large territory about. In the immediate region examined the quartz-porphyrites are most common; their trend is often with that of the slates, though occasionally across it. They occur, in some instances, in proximity to, if not immediately next, the ore-bearing slates; in others they are at a distance. Their influence upon the ore bodies could not be ascertained from the few shallow prospects existing.

The ore bodies are the result of the mineralization of certain layers in the zones of slates or schists. Quartzitic, micaceous, and calcareous beds all carry ore in greater or less amount. If the zone of slates is narrow—1 to 2 feet—the entire width may be more or less mineralized; in this case the walls are granite. In the wider belts the ore shows in considerable deposits in some of the layers, in others in much less amount; in others still it may be altogether wanting.

Evidences of fissures were slight, though in some instances the rocks were contorted to a considerable degree.

The ores are argentiferous galena, and antimonial and arsenical sulphides, also silver-bearing, probably. They are more or less altered near the surface. Their value varies greatly, but in the assorted ore it is sufficient to warrant shipping by pack train.

THE YELLOW JACKET DISTRICT.

This district lies southwest of Salmon City, on Yellow Jacket Creek, a tributary of Camas, which in turn enters the Middle Fork of Salmon River. It is in the midst of mountains of great ruggedness and is approached by trail from Challis and Salmon City, each about 60 miles distant. From the latter place, however, a good road has just been completed via Leesburg to the mouth of Fourth of July Creek, 12 miles from the mining camp. The outward communication from Salmon City is by wagon road 75 miles to Red Rock, on the Utah Northern Railroad; from Challis by wagon road 75 miles to Ketchum, the terminus of a branch of the Oregon Short Line. The altitudes above sea-level along the Salmon City road west are given below.

Altitudes between Salmon City and the Yellow Jacket district, Idaho.

	Feet.
Salmon City, about	4,000
Divide between the waters of the main Salmon and Big Creek	8,500
Leesburg	6,550
Big Creek at the mouth of Napias	5,500
Mouth of Fourth of July Canyon	5,900
Divide between the waters of Big Creek and Yellow Jacket Creek	7,800
Yellow Jacket Camp	5,875

The readings, which are from an aneroid, if out, are a little too high. The Yellow Jacket camp is 10 to 15 miles below the head of the basin drained by Yellow Jacket Creek and its upper tributaries. The gorge of the main creek at the camp is about one-fourth of a mile wide, closing to 50 or 75 yards just above, with occasional narrow openings still higher. Below the camp the width of one-fourth to one-half of a mile is maintained for at least a mile or two, and along this portion are gold placers in preparation for being worked. The sides of the gorge are comparatively steep, and the country about is well timbered. Two or three gulches enter the Yellow Jacket at the camp, one from the south, the others from the north, the latter short, though reaching well toward the summit of the ridge. The two great lode properties are those of the Columbia Consolidated Gold Mining Company and the Yellow Jacket Mining Company. Other claims are located but undeveloped. The Columbia property is northwest of the main stream, in the divide between this and its tributary, Hoodoo Creek. The east side of this divide is that on which present exploitation is conducted. It presents a steep, slightly indented slope, advantageous for the opening of mines. The Yellow Jacket lode is between 1 and 2 miles northeast of the Columbia and also north of the main stream. It is located well up the slope of a steep hill, the southern terminal of a spur from a massive ridge in the curve of Yellow Jacket Creek. This property is separated from the Columbia by the two lateral gulches mentioned above. The mining region in its entirety extends to the east for a mile or two up Yellow Jacket Creek in the hills on either side, and to the west about the heads of Hoodoo, Lake, and Wilson creeks. West of Wilson Creek is the divide separating it from the Middle Fork of the Salmon. South and southeast of the Yellow Jacket camp the ridge separating it from Camas Creek and its tributary, Silver Creek, is high and rugged, its topographic configuration being most irregular. The ridge is the site of great volcanic activity in past time.

Yellow Jacket Creek is a small stream, but the water is sufficient for milling and other purposes in connection with vein mining. For the placers it must be utilized chiefly in the early part of the season.

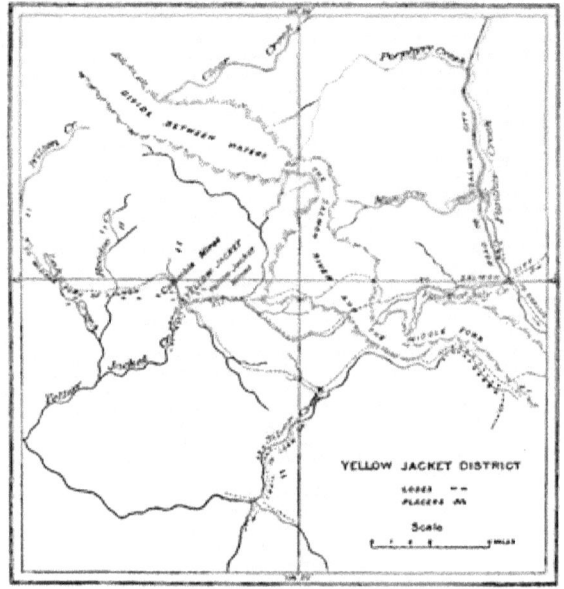

FIG. 40.—Sketch-map of Yellow Jacket district, by F. D. Howe.

The geology of the Yellow Jacket region was but generally examined in the reconnaissance. The district lies well within the area of crystalline schists, but local exposures of the gray granite so common in Idaho occur in the ranges about, some of them probably of considerable area. It is reported, for instance, in extended outcrop on the head of the main fork of Yellow Jacket Creek, 10 to 15 miles above the camp, in the divide between the waters of the Salmon and the Middle Fork. Abundant débris from this source shows along the creek bottom. It is also said to occur a few miles west of the district, along and beyond the Middle Fork, a statement borne out by the appearance of the topography in that direction. The prevailing rocks, however, are the crystalline schists, quartzitic and micaceous chiefly, but with occasional bands more or less calcareous. The quartzitic variety predominates, and in one instance, in a layer of typical quartzite, distinct ripple-marks were observed on the stratification planes. The series is thin bedded throughout.

The eruptive rocks include rhyolite, trachyte, andesite, quartz-porphyry, mica diorite, syenite, aphanitic syenite and minette (lamprophyre), and diabase. Of these the quartz-porphyries, rhyolites, and andesites occur in the heaviest and apparently most irregular bodies; the others, usually as narrower dikes. The quartz-porphyry, and the aphanitic syenite and minette (lamprophyres), called "syenite" by the miners, are most commonly associated with the metalliferous veins. The lamprophyres, according to the observations of Mr. F. D. Howe, manager of the Columbia Consolidated Gold Mining Company, when present and in contact with the veins, are usually their foot wall, while the hanging wall may consist of quartzite, quartz-porphyry, or some other one of the eruptives. The lamprophyre dikes are considered as indicative of the near presence of an ore body, so persistent is their occurrence in this connection; at least this is the case on the Columbia Hill. On the Yellow Jacket Hill the lamprophyres do not seem to occur so often nor in such intimate connection with the veins; indeed, the veins themselves differ in the two hills. The diabase was observed only on the Yellow Jacket Hill. The andesite occurred on the divide between the Fourth of July and Yellow Jacket drainage, while the rhyolites were most conspicuous 1 to 2 miles east of the camp and on Fourth of July Creek. Quartz-porphyries in heavy masses occurred both at a distance and on the crest of the ridge just above the Columbia mines. Occasionally a dike appears to be interbanded with the schists, but in most cases they cross the dips. The influence of the eruptives upon the mineralization of the region is undetermined.

The entire region between Panther and Camas creeks, which includes the Yellow Jacket mining district, has been in former times one of great dynamic disturbances; folds, flexures, and faults, in addition to the intrusion of the eruptives, occur. The structure is therefore difficult to determine, and the difficulty is increased by the similarity of the schists from the base to the summit, rendering identification of horizons impossible. In general the structure as suggested by strikes seems to have been developed on lines varying from N. 30° E. to N. 20° W., with local divergences to N. 60° E. and N. 60° W. This is evident both in the divide between Panther and Yellow Jacket creeks and the country to the west, to and beyond the Yellow Jacket mining camp. Moreover, it is on these lines that the majority of the eruptives have cut through, although there are doubtless instances in which the dikes cross the schists at greater or less angles with their strike. Again, in some of the folds, on both small and large scale, slates and dikes together appear to have swerved from the normal or more usual trend.

The dips in the Panther-Yellow Jacket divide, along the route traveled, are nearly all steep. On the eastern side the beds more commonly incline to the east when departing from the vertical, and this direction becomes pronounced along Big Creek between Fourth of July and Napias, where dips as low as 20° are sometimes encountered, from 20° to 35° being quite common. On the western side of the divide a vertical position is maintained by much of the series, but in instances of departure from this a westerly dip is perhaps most usual, though a steep easterly inclination is occasionally met with. The westerly dip becomes more pronounced on approaching the vicinity of the Yellow Jacket camp, notwithstanding that in this immediate region the strata have been extensively fractured and crumpled, with consequent variation in both strike and dip. West of the Yellow Jacket camp the structure lines of the formations are unknown.

In general it appears from the foregoing observations that the high range forming the Panther-Yellow Jacket divide marks the axis of an anticline, and that on the west side, especially, the rocks have been locally fractured and bent, affording an opportunity for the locus of mineralization in the hills about the Yellow Jacket camp.

In the Columbia Hill the schists are particularly crumpled and fractured. There are several lines of structure, all bearing upon the geology of the ore bodies. Three, N. 20° W., N. 15° E., and N. 30° E., are strikes of the schists, representing flexures in their trend. Occasionally these are also the courses of joint planes. Two other lines of structure, N. 60° E. and N. 15° to 25° W., prevail throughout the region, the directions of joint planes and perhaps also of strikes of the schists. These directions for the joint planes become especially significant when it is known that they are also the directions of the two chief systems of ore deposits of this hill, the veins of a N. 60° E. trend forming the principal system. The observed dips of the locality were all to the west, varying from 10° to 60° and even 90°.

In the Yellow Jacket Hill, a mile or two northeast of the Columbia, the strike of the schists is N. 50° to 60° E., and this is here the strike of the main vein of the property also. Schists and vein together have an average dip of 33° W., locally increasing to 45°, 50°, or even 70° or 90°, but the steepest dips are rare. Faults doubtless occur in both the Columbia and Yellow Jacket hills, but their relations to the ore bodies, eruptives, and country rock, in strike and dip, were not determined.

The principal system of metalliferous veins in the Columbia Hill— that having the N. 60° E. trend—appears to be a series of mineralized zones of highly fractured material, in width from a few feet up to possibly 75 or 100; 50 was the maximum observed by the writer. Their dip varies between 30° and 70° W. The foot walls are frequently aphanitic syenite or minette (lamprophyre; local term, "syenite"), while the hanging wall may be either quartzite, quartz-porphyry, or some other eruptive, occasionally even the syenitic rock. The eruptives in general parallel the veins in strike, and, in some instances at least, the quartzites also seem to do so.

The fractured zones are clearly defined, but it was not possible in the time available for the examination to determine satisfactorily whether they were at all points parallel with the line of strike of the schists or in part independent of and divergent from this. Again, it was equally impossible to determine in the yet comparatively shallow openings whether the zones, when coincident with the schists in strike, dipped with them or at a greater or less angle. Whichever may be the case, it will not alter the character of the deposit. The vein material is a breccia of fractured schist, in one or two instances including pieces of the accompanying lamprophyre ("syenite"). The breccia is bound together with siliceous matter (quartz), and the filling completed with the mineralization of the zone. From the included lamprophyre it would seem that at least the fissures it filled existed prior to the fracturing which resulted in the brecciated, mineralized zones. In regard to the other eruptives, the time relations are obscure.

The metallic contents of the fractured zones are carried not only in the interstitial quartz but often in the fragments of the schists and slates themselves, impregnation of these having taken place to considerable depths, though perhaps not to the complete replacement of the central portion. The exterior of such fragments under atmospheric or other oxidizing agencies often appears rusty, from the partial alteration of the sulphides to oxides or carbonates. The deposition of the metallic contents has been more or less unequal throughout the lodes.

The system of veins on the Columbia property having a N. 15° to 25° W. trend is less prospected than that just described. Such veins have in a few instances been observed coming into one or another of those of the N. 60° E. system, but their extent in the line of their trend is undetermined. Neither could it be observed at the time of the examination whether at all points these were confined to the areas in which a N. 15° to 25° W. strike prevailed for the schists; whether, in fact, they did not, as suggested in one or two instances, form lateral "feeders" to or offshoots from the N. 60° E. system, on the stratification or other planes, or even as fissures. It is significant that joint planes having a direction of N. 25° W. are of frequent occurrence throughout the property of the Columbia Company.

The ores of the Columbia mines are chiefly copper sulphides carrying gold and silver and ranging in value up to $150 per ton, with concentrates increasing in richness according to the completeness of the operation. An average ore is stated to be about $30, while $40 to $50 is not infrequent. The sulphides near the surface are more or less oxidized, or altered to carbonates, affording a free milling ore; and it is stated that there is more or less free gold in the unaltered portions also.

In the hill embracing the property of the Yellow Jacket Gold Mining Company there appears to be, so far as shown by the present state of exploitation, but a single vein, or at most two, if we except a number of small stringers in their vicinity. The country is a quartzitic schist in layers of varying thickness and hardness. The veins are of quartz, and lie with the schists, striking N. 60° E., and dipping northwest on an average 33°, increasing at points to 45°, 60°, or even 90°. But a single instance of a breccia was observed in the superficial examination given the mine; this consisted of included fragments of country rock in a calcite cement. The highly brecciated condition prevailing in the Columbia mines appears to be absent in the Yellow Jacket. The vein has been traced but a short distance beyond the Yellow Jacket Hill. The thickness of the vein is said to reach 40 feet locally, but the maximum seen by the writer was about 15 feet. The quartz is said to occur in lenticular bodies of varying dimensions. Occasionally eruptive dikes appear in the mine, chiefly of aphanitic syenite or "minette" (lamprophyre) and diabase. Their relations to the ore bodies were not studied.

The ore of the Yellow Jacket mine, or at least of that portion of it visited by the writer, is a free-milling, auriferous quartz, with one or two small local bodies of hematite and an occasional but rare copper stain. This difference from the ore of the Columbia mine is as marked as the difference in the manner of occurrence and character of the veins themselves. The mass of the ore of the Yellow Jacket mine is said to run from $7 to $30 per ton, with local values much higher. An ore of common occurrence is one of $18 or $19.

The formation about the summit of the Yellow Jacket Hill has undergone considerable disintegration, and the surface is covered with débris which is reported as carrying free gold from a trace up to $7 per ton. This may be milled at a profit.

THE WOOD RIVER DISTRICT.

This district embraces an area 15 to 25 miles square on upper Wood River and adjoining streams, just within the southern edge of the mountain mass of Idaho. The greater portion of the district lies within the confines of the Wood River drainage. This has formed a topographic depression ridged with spurs of the inclosing ranges, the intervening gulches being sharply eroded to depths of 1,500 to 3,000 feet. The central valley, one-fourth to one-half mile wide, ranges in altitude from 5,000 feet at its lower end (Bellevue) to 6,000 feet a little above Ketchum, rising still more rapidly beyond. The periphery has an elevation of 9,000 to 10,000 feet, a single peak, Mount Hyndman, 15 to 20 miles southeast of Ketchum, reaching 12,000 feet, the highest in the State.

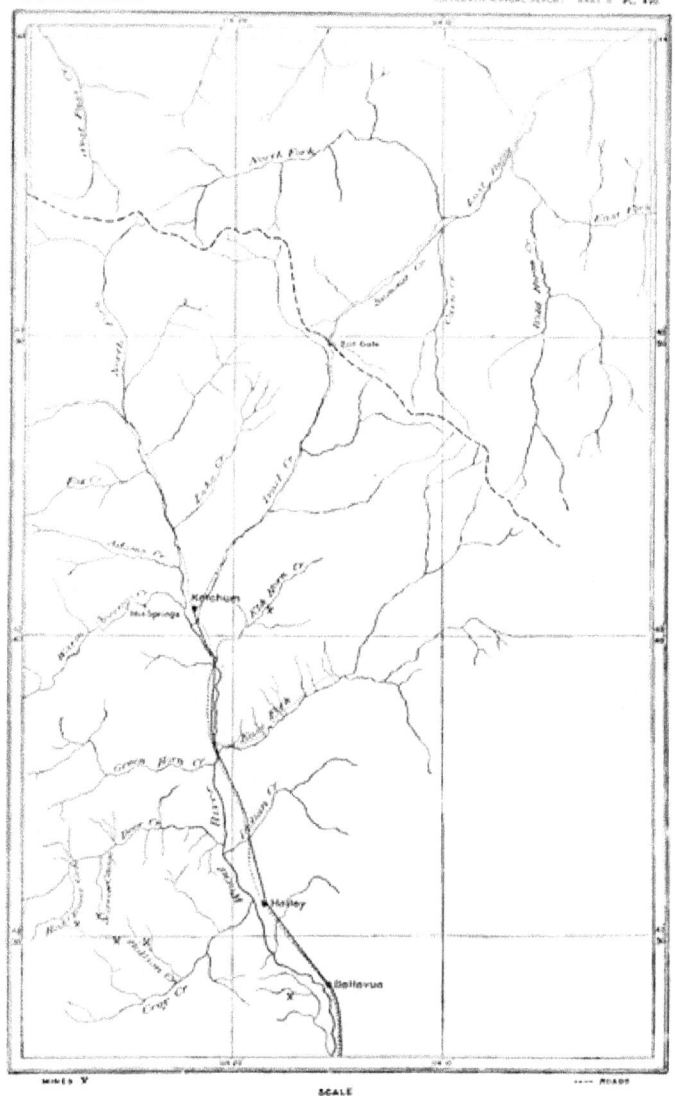

SKETCH MAP OF WOOD RIVER DISTRICT.

The ranges and their spurs are somewhat rugged, and their configuration varies according as the rocks entering into their composition are quartzites, limestones, or shales. Most of the streams are comparatively small, but Wood River has a large volume of water. The bottom lands along the main stream are cultivated and supply the mining camps; the hills are timbered in varying amount, though the mines are not always in close proximity to the heavier bodies. Hailey and Ketchum, 12 miles apart, are the important towns of the district, and are connected by a branch with the Oregon Short Line at Shoshone.

A résumé of the general geology for the district shows a base of gray granite; resting upon this, in one locality, dark and light colored quartzites with calcareous beds; in another, a heavy mass of pink quartzite, varying in thickness up to 500 feet; and in a third, a great series of black or dark-gray calcareous shales. Intercalated somewhere in this series, probably between the gray shales last mentioned and the dark quartzitic series, are from 300 to 500 feet of heavy-bedded limestone, resembling somewhat the sub-Carboniferous of the Rocky Mountains. The arrangement suggested as most probable, in the light of the evidence at hand, is granite, heavy pink quartzite (Cambrian), dark and light quartzitic series (at least post-Cambrian), the limestone resembling the sub-Carboniferous, and the gray calcareous shales (Carboniferous). All these are cut by eruptives—hornblende-andesite and hornblende-mica-andesite. The upper, calcareous, shaly division carries most of the mines of the Wood River district proper, but in the outlying portions the dark quartzite series and the granite also contain mineral deposits of one kind or another.

Throughout the Wood River region proper there is a general N. 25° to 40° W. strike with a southwest dip, usually between 30° and 60°; but local variations occur, sometimes involving a region of considerable importance. The southwest dip is that of the half of a probable anticline, the eastern half being in the Lost River country, and the axis along the divide between this and Wood River.

Faults are of frequent occurrence throughout the region. They are of many directions and of varied inclination and throw. The amount of throw is often obscured by the similarity of the strata on either side of the fracture plane and by the great thickness of the series opposed. Following are the types observed, which are probably representative for the region.

One, the plane of which is horizontal, displacing a dipping ore body 50 feet or more; the upper part of the vein is carried in the direction of the dip, both ends of it being broken over, with a dragging of the ore along the plane of the fault. Faults inclined 25° to 50° also occur, their strikes often divergent; such faults are known to have drained the upper levels of a mine when struck at a lower. These faults may sometimes be traced at the surface by a break in the topography, a line of gentle depression or otherwise. It is often significantly remarked by the miners of various districts that the ore follows the line of a gulch. The position of a gulch is often determined by a fracture-line, and it is along a fracture-plane that mineral solutions have found their way; hence the coincidence. Fracturing has also taken place with a minimum of displacement, at times no more than a crushing of a particular zone with a slight slipping of fragments upon one another or upon the walls. This was observed both coincident with the stratification and across it. Slickensides are observed nearly everywhere that there has been movement between the rocks on the two sides of the faults, the grooves lying in many directions. In addition to the above recognized faults there are many points along Wood River and its tributaries where the series of calcareous shales opposes the underlying heavy-bedded gray limestone, which is indicative of considerable throws. The faults of the Wood River region are among the most important features requiring examination. Flexures also occur, but study of their details was not attempted.

Mineralization of the veins in the Wood River district has apparently taken place in a manner wholly independent of the eruptives. The region has at different periods been one of great fracturing, on both a large and a small scale, and passages have thus been afforded, in some instances for the intrusion of dikes and other irregular bodies of eruptives, in others for the flow of mineral-bearing solutions. These solutions have not only filled the fractured zones themselves but have oftentimes found their way into the more solid beds of limestone adjoining, replacing them to such a degree that they too became of economic value.

The mines of the Wood River region that have thus far been most productive lie within the periphery of the drainage basin itself. A single exception, perhaps, is the Camas No. 2, near Doniphan, just beyond the southwestern border. The position of the mines may be isolated, or there may be certain localities where the deposition of ores has been especially active, possibly through an easier access afforded the mineral-bearing solutions. For lack of time it was impossible to do more than (under the guidance of two or three of the mining men acquainted with the district) to visit a few localities which together would afford a general idea of the occurrence of the ores and of their character.

On both sides of Wood River, fracturing, faulting, and folding were suggested in the distribution of the strata and their many variations in strike and dip. For the region east of the main valley the general anticlinal structure described for the southeastern portion of the Sawtooth Range prevails. The strike most common is N. 25° to 40° W., the dip usually southwest. Faults of greater or less extent and magnitude may be seen both in the mines and upon the surface, and folds were discerned from a distance by the writer. The area of examination, however, was very small. On the west side, the region about the head of Deer Creek and its tributaries, extending over to Bullion Creek and the gulches west, received most attention. It is 9 to 10 miles west of Hailey, and 2,000 to 2,500 feet higher. The geology along lower Deer Creek is obscure and subject to sharp changes in both structure and stratigraphy. South of the creek limestones bearing a close resemblance to the heavy-bedded gray variety of the district appear in the bluffs in one or more flexures, while eruptives form a portion at least of the higher hills beyond. On the north, for the first mile, an eruptive closely resembling the hornblende-andesite higher up Wood River occupies the interval opposite the flexed limestones on the south. The eruptive finally gives way to the dark quartzitic series; this in a mile or two to a heavy-bedded blue limestone, which also now appears on the south side of the valley. In about 200 yards the limestone is succeeded by the black and gray calcareous shales, here more calcareous than at many points in the district. The general strike of this series is across the creek, the dip probably upstream, or west.

At the Warm Springs, which are about 4 miles above the mouth of Deer Creek, the slates are succeeded by gray granite; it is impossible to state whether in natural sequence or by faulting, but the latter is suspected. The granite extends for 3½ miles along the creek, the foliation planes having the general strike of the country, N. 25° W., and dipping upstream, westward. The granite is overlain by 500 feet of white quartzite, which, however, disappears within a mile or two south of the creek, the manner of disappearance not being determined. This is overlain by the black and gray calcareous shales and limestones, the ore-bearing series of the Wood River region, which, where the quartzite has disappeared, at the head of Narrow Gauge and Bullion gulches, come directly down on the granite. The dip of the entire section, from granite up, is westward until at the head of Deer Creek and of its tributary gulch, Red Cloud, it gradually changes to the southwest and south, the strike of the slates changing accordingly. This departure in the strike and dip is particularly conspicuous in the high ridge separating Red Cloud Gulch from the gulches to the south in which the Red Elephant and Bullion group of mines are located. For all this, however, it is a local feature, and the normal strike of N. 25° W. prevails over much of the area included within this southwestern corner of the Wood River district.

The hill in which the Red Cloud, Red Elephant, Bullion, and French groups of mines are opened is one of marked fracturing as well as folding, both on a large and small scale. The fractures occur at angles divergent with the strikes and dips, and also along their planes, and are locally so complex as completely to obscure the bedding. No clearly defined system of fracturing has thus far been worked out for the hill, but that most prominent has a N. 25° W. direction. This also carries several of the more important mines. The greatest irregularity appears to be in the northern face of the hill. In the southern, or at least in the spurs on this side, the strata seems to have regained the strike normal for the district in general, N. 25° W., the dip being uniformly westward, 30° to 60°, or locally even nearer the vertical. The precise connection between the structure of the northern and southern faces of the hill was not determined.

The hill just described is on the very edge of the great mountain mass of Idaho, but few foothills intervening between it and the lava plains of the Camas and Snake valleys. Neither the region at the very head of Deer Creek and its tributary, Red Cloud Gulch, nor the country to the west of this was visited. It is said that granite lies beyond the Wood River drainage, in this direction, and the observations of the writer from a distant point to the south confirm this in part. It is quite possible that such succession of granites west of the Deer Creek shales and slates is brought about by a bending up of the strata to the west, making a syncline somewhere between the Deer Creek mines and the granite. Faulting is the alternative.

The ore deposits of the Wood River region, so far as observed by the writer, are of three classes: those occupying fractured zones in the great body of calcareous shales and limestone; those occurring in beds of limestone; and typical fissure veins. There are instances, however, in which these classes seem to occur in connection one with another.

The fractured zones are more or less prominent throughout the district. Such zones have the appearance of a breccia of the country rock, with the interstices filled with quartz, not infrequently supplemented by calcite in minor amounts, the whole more or less impregnated with the metallic contents. The width of such zones varies from a foot or two up to 15 or 20 feet, and may be even greater locally. The zones may coincide with a particular bed of limestone or may cut across both strike and dip at various angles of divergence. When this divergence is slight it is difficult to distinguish between an ore-bearing bed and an independent zone of fracture, except through the breccia structure. What are called foot and hanging walls are also recognized in veins of this kind. Where the fractured zone crosses the strata, occasionally a lateral ore-shoot follows a bedding plane for a short distance, given off from the main channel of mineralization; indeed, mineralization along bedding planes, however slight, is considered an indication of the near approach to a vein.

The fractured zones may have resulted from simple crushing with a minimum of movement; or, as is apparently often the case, the movement may have been considerable. The creation of the spaces now occupied by interstitial filling may have been by actual separation of the fragments; or they may have been merely a network of joints and cross-joints along which solvents easily passed, carrying away portions of the lime and so creating interstices which were subsequently filled with the mineralizing solutions; or, perhaps, even the calcareous material of the breccia itself was gradually replaced by the metallic compounds. When considerable movement took place in the zone of fracture, fissures were often created within the boundaries of the zone, the deposits filling them assuming the character of local fissure veins. It is possible that in such cases the fissuring may have been wholly secondary to the primary fracturing—cracks opening within the already partly mineralized body of rock. In other cases the original movement may have resulted in a fissure, the present brecciation being due to included fragments of the country rock, themselves, perhaps, minutely fractured.

In the mineralization of the fractured, ore-bearing zones, the metallic sulphides occur distributed through the interstitial filling, or accumulated in thin layers along partings in the vein material, or, again, in massive bodies of ore, 1 to 20 feet thick, on one or the other of the walls, in the interior of the zone, or occupying its entire width. Many instances of the replacement of limestone within the fractured zone by metallic compounds occur. Such might also have been the case with the heavier bodies of ore. On the other hand, the heavier bodies may have been deposited from solution in the fissures opened in the fractured zone. Oftentimes, where there is a vein of nearly solid ore, or at least a body of highly mineralized rock in a fractured zone, the portions of the zone on either side, if the body is in the interior, may also be mineralized in paying amount and the adjacent country itself may carry a small percentage of ore.

Of the class of ore deposits occurring in beds of limestone a single instance was observed by the writer, but this form of deposit may not be uncommon in the Wood River district, where like conditions are maintained over such an extended area. The deposit bore considerable resemblance to the deposits at Leadville, in the "blue limestone" of sub-Carboniferous age. It was not, however, accompanied by eruptives, and this absence of eruptives in connection with ore bodies prevails in the Wood River region; the source of the mineralization is therefore quite different in the two localities. In the deposit observed by the writer there is a mineralized zone of limestone with an included vein of ore. Where seen the total width of the metalliferous zone was 5 or 6 feet, and that of the included vein 3 inches to $1\frac{1}{2}$ feet. Both are said to have reached greater width in other portions of the mine, 12 feet of galena being reported at one point. The country rock is the dark-gray calcareous shale and limestone, the gangue, quartz with a little calcite, and here and there portions of shale. The ore bodies in veins of this nature are said to be nearly independent of one another, and appear to have originated in the mineralization of larger or smaller portions of limestone which were within easiest access of the mineral-bearing solutions. It is quite reasonable to suppose that in a region of this kind in the folding which the strata have undergone bodies of limestone, parts of an entire bed, may have been reticulated with cracks without displacement of the fragments, and that through such bodies solutions found an easy passage. In a region of such fracturing and so many channels for the flow of mineral solutions, any of the limestones may become locally ore-bearing. For the preference of the limestones of the shaly series over the massive gray limestone of the region, no explanation was arrived at; the suggestion occurs that it may be due to original composition.

The class of veins which may be considered as having had their origin in fissures—fissure veins—occurs both in the great calcareous slate and limestone series of the Wood River district and in the granite just beyond. In the slates they are doubtless numerous. They are here distinguished by their strike and dip, which is at variance with that of the slaty beds, by slickensides, by selvages of clay, by occasional included fragments of the country, and by the character of the filling, which is a heavy deposit of quartz in one or more layers, with or without accessory minerals like calcite, the metallic contents being deposited in masses along the planes or disseminated throughout the ledge matter. Moreover, as the vein crosses the stratification of a series of thin-bedded rocks, the nature of the country is constantly changing, both on the course of the vein and in depth. Original fissure veins in slates contain more or less débris from the country, but are usually easily distinguished from the zones of fracture and their included local fissures.

Near Doniphan, on the head of the east fork of Camp Creek, southwest of the great area of calcareous slates, and in the region of granite, is a prominent quartz ledge having a general course of N. 30° to 35° W. and extending with interruptions for 3 or 4 miles across the country, Doniphan being near its southeastern terminus. The granite is somewhat foliated, and, so far as observed, the planes have a prevailing strike N. 25° W. and a dip southwest. The ledge presents some minor curvatures of strike, and the dip also varies a little from point to point, but is in general about 45° NE. The vein doubtless belongs to the class of true fissures—a view strengthened by slickensides, clay selvages, and fragments of the country in the ledge matter. There are several locations upon the ledge, but the Camas No. 2, at the southeastern end, is the only mine of present importance and is said to have been a good producer in earlier times. The stopes would confirm this. It is now about being reopened. The width of the ledge varies, the maximum observed being between 6 and 10 feet. The quartz is banded, coarsely crystalline and showing little mineral, or fine-grained and mineralized to considerable amount. At a single point a bit of limestone resembling that of the shaly series was discovered adjacent to the ledge, but its presence was inexplicable. Its appearance, however, was that of an intimate part of the rock, not an included fragment. The ore occurs in shoots of greater or less size, the mineral contents being either disseminated in particles throughout the entire width of the ledge, concentrated in bunches on the walls or in the interior, or confined to certain benches.

The ores of the Wood River district are lead-silver and gold. The former is practically confined to the great series of calcareous shales, the latter to the granite.

Associated with the galena of the first class of ores are blende, pyrite, arsenical pyrites, gray copper, erubescite, and occasionally native silver. Blende and common pyrite are the most widely occurring and abundant associates; the others are comparatively rare. Carbonates usually occur near the surface. The galena is fine to coarsely crystalline, and occurs disseminated, bunched, or in extended bodies. It often shows striation resulting from twinning. The blende occurs either through the galena or in distinct portions of the vein with but slight admixture of galena. The pyrite may occur in the same manner as the blende, but is of wider and more universal distribution than any of the other minerals, and moreover is frequently found in the country adjacent to the ore bodies. No examination as to the contents of the ores was attempted, but they are said to run as high as 160 ounces of silver and 70 per cent of lead to the ton, with an average considerably below this.

The gold ores occur in the veins in granite. They are in part free-milling, in part smelting. The minerals associated with the gold are pyrite and chalcopyrite, with their alteration products, and a very small amount of galena, probably argentiferous, as silver is found in such ores on examination. The milling ores are said to run from $8 to $15 per ton in free gold, the assays indicating a value of $26 in this metal.

THE SILVER CITY DISTRICT.

This is located on the southwestern slope of the Owyhee Range, about the head waters of the Jordan River, a tributary of the Owyhee; the district extends, however, to the heads of other streams in the vicinity, and altogether occupies an area 8 or 10 miles square. The two leading towns are Silver City, at the head of Jordan Creek, and De Lamar, 7 miles below, to the west. The Jordan Valley and its tributaries are sharply eroded to depths of 2,000 to 3,000 feet below the crests of the range and its spurs. The water supply is light. Timber is scarce. The district is 50 miles from the railroad, Nampa on the Oregon Short Line being its station. An excellent wagon road over the mountain and across the Snake Valley connects the two points.

The Owyhee Range, in which are situated the two important mining camps of Silver City and De Lamar, is primarily a granite range, in later times cut by rhyolite, diabase, and basalt.

The two leading mines of the district are the De Lamar at De Lamar, and the Black Jack, 1½ miles west of Silver City. Both are on clearly defined fissure veins in rhyolite. There is, however, in the Black Jack an associated rock, diabase, and on the mountain side below this mine basalt occurs. The rhyolites are several miles in extent, but whether the two mines are on the same body was not ascertained. The rhyolite in a decomposed state is sometimes mistaken for granite by the miners. The diabase, which is called "trachyte" by the miners, occurs locally on one side of the vein in the Black Jack mine, and in the Trade Dollar, to the east of the Black Jack, it is said to form both walls. This rock is younger than the rhyolite with which it is associated.

The veins trend between N. 10° and 60° W., varying locally; the prevailing dip, which is steep, is west or south. Fracture planes cross the veins at greater or less angles, but examination was not in sufficient detail to determine a system. The width of the veins is from an inch or two up to 10 feet or more. The vein matter is quartz, banded or massive, coarsely or finely crystallized. The quartz locally shows a brecciated structure, coarse or fine, indicating movements since the formation of the veins; the fragments are often somewhat abraded and surrounded by a clay gouge. Not infrequently, also, included

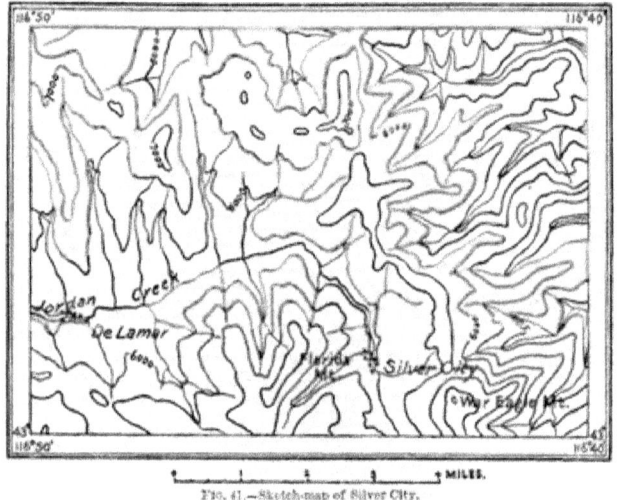

Fig. 41.—Sketch-map of Silver City.

fragments of the country rock are found, either as horses or in small, angular pieces. Talc, so called, is of common occurrence in certain localities in the mines, even in the interior of the ledge. Besides this, there is the usual selvage on the walls, attaining locally a considerable thickness.

The ores are gold and silver. Both metals occur native, but the silver is found in large proportion as argentite. The chloride is also present. Pyrite, auriferous and nonauriferous, occurs. The talc in the veins is a remarkably productive class of ore, carrying a large amount of argentite. The ores are generally of very high grade.

PLACERS.

The mills of the companies are well equipped and are kept running night and day a greater part of the year.

The placers on the route of reconnaissance are chiefly gravel bars along stream bottoms. From reports some of them have yielded fabulously in the past, but at present work is prosecuted only by a few scattered Chinese laborers. The placer gravels encountered were all on streams that had their sources in the gray granite or in the crystalline schists which lie upon this. Indeed there are apparently but few streams rising in these terranes that do not at some point carry pay gravel or have not already produced more or less gold. Among the more notable instances are the bars of the main Boise and its tributaries—South Fork below Rocky Bar, and Middle Fork, below Atlanta; Loon Creek, entering the Middle Fork of the Salmon; and Yellow Jacket Creek below the mining camp. Leesburg, on Napias Creek, has been the center of a great placer district in early days, and at California Bar, 3 miles below, a company is now preparing to work a supposed extensive placer.

Different from the foregoing class of placers is that on Kirtley Creek, 8 miles east of Salmon City, within 2 miles of the base of the Continental Divide. There is here, indeed, as along several of the streams heading in this range, the usual river and creek gravel bars, carrying a greater or less amount of gold, but in the bluffs of the creek there is also an extensive placer probably of Tertiary age, which is now being opened. The general feature of this deposit is, briefly, a succession of Tertiary gravels and sands, which possibly correspond with beds farther out in the Salmon Valley that from paleobotanic evidence are Eocene or Miocene. The material was derived from the neighboring range, and formed a shore deposit of the intermontane lake. Their north-and-south extent was not learned, but it may be several miles. The gravels may not everywhere be auriferous, however.

The deposit as exposed consists, from base up, first, of a series of conglomerates and sandstones, about equally divided, cut to a depth of 15 feet, and found to be auriferous—in paying amount, it is said, under a sufficient supply of water. Upon this series is a very white, thin-bedded sandstone but a few feet in thickness, succeeded by two other layers, of white and yellow sandstones, 5 and 3 feet thick respectively, which are leaf-bearing; these are called by the miners bed-rock, as they form the foundation for the overlying gravels, which are unconformable and said to be fairly rich, especially near the bottom. The overlying gravels are fully 20 feet thick when present in their entirety, and are in turn succeeded by 30 feet of interbedded sandstone and conglomerate, the sandstones predominating, and all carrying more or less gold. This completes the succession of beds of supposed Tertiary age. The entire series, both above and below the line of nonconformity, is flexed, a westerly dip of 15° to 45° prevailing. This bending of the strata indicates that if the portion above the break is not of the same age as that below, it is at least earlier than Pleistocene.

Resting upon any of the foregoing beds, according to their exposure in the past, is a Recent gravel, derived in part from that below, in part, perhaps, from the present mountain slopes. This gravel is said to be the richest of all, the result of a natural concentration of the earlier auriferous gravels. The gravel, both Tertiary and Recent, is of quartzite or quartzite-schist débris; the Tertiary portion is usually tightly cemented and somewhat difficult to hydraulic. The gold contents were not ascertained, but elaborate preparations are being made for mining.

The present supply of water is from storage reservoirs in the mountains, and is insufficient for constant work. It is proposed to take water from the Lemhi River, which would probably obviate this inconvenience.

COAL.

Two coal areas were encountered on the reconnaissance, one about Salmon City, the other at Horseshoe Bend on the Payette, 28 miles north of Boise. Both areas are probably Tertiary—Eocene or Miocene—confirmatory plant remains having been found in the beds at Salmon City and the series on the Payette bearing a close general resemblance to these in composition, in manner of occurrence, and in folding. The coal occurs in small seams in sandstones, shale or conglomerate lying within a few feet.

SALMON CITY VALLEY.

The only opening in this valley is a drift 20 feet in length in the east bluff of the Salmon River a mile below the city. There are here several thin streaks of lignite—the thickest, 6 inches—in 5 feet of carbonaceous slate and sandstone. The lignite has a dead-brown appearance, and its woody structure is plainly visible; the ash is high, as are probably also the water contents. It is valueless so far as exposed. The measures which carry this lignite underlie a large portion of the valley about Salmon City, and at several points show thin beds of carbonaceous shale, indicating the former presence of plant life and a tendency to the formation of coal; but the conditions seem not to have been those requisite for the development of a coal of economic value.

HORSESHOE BEND, VALLEY OF THE PAYETTE.

The valley of the Payette at Horseshoe Bend has a northeast trend, is about 9 miles long by 3 across, and is divided midway its length by the axis of an anticlinal arch having a general northwest direction. Coal occurs in both the northern and southern halves of the valley. The northern prospects were not visited, being reported but slightly opened and in unsatisfactory condition for examination. The only southern opening at present is a surface cut and short tunnel on what is known as the Rob vein, of which the following is said to be a section, though only the upper coal was visible at the time of visit:

Section of the Rob coal vein, Horseshoe Bend, valley of the Payette.

Sandstone and shale	roof.
Coal	1 ft. 6 in.
Slate	1 ft. 0 in.
Coal	0 ft. 10 in.
Slate	1 ft. 0 in.
Coal	0 ft. 10 in.
Clay	floor.

The strike of the vein at the exposure is northwest; the dip, SW. 5° to 10°. A short distance south an eruptive a few hundred feet wide cuts the coal. Beyond this the measures again appear, but are of limited extent, owing to the proximity of the granite of the Boise Range, or of a second eruptive. Northwest the coal bed may extend for a considerable distance before reaching the edge of the valley, the western wall of which is granite.

The coal is black and comparatively hard. Following is the analysis of a sample taken from a small pile on the dump, which is said to have come from a tunnel inaccessible at the time of visit.

Analysis of coal from Horseshoe Bend, valley of the Payette.

Moisture	4.84
Volatile matter	36.23
Fixed carbon	54.55
Ash	4.38
Total	100.00

The excellent quality of the coal in the present instance may be explained by its proximity to an eruptive. It is questionable whether this grade would be maintained over the entire valley.

AGRICULTURE.

In crossing and recrossing the State the following facts in regard to agricultural possibilities were impressed upon the writer: Much of Idaho is a rugged mass of mountains, and but little area is available for farming in comparison with that of many other Western States. The intermontane valleys are few, but are very fertile and well adapted to hay, grain, vegetables, and fruits. Berries, apples, peaches, plums, and prunes, all, in one locality or another, attain remarkable perfection. In the Boise Valley and the country adjacent the prune, plum, and peach industries have been successfully started and promise enormous advances in the next few years. The apple industry is also large, but in this there is apparently a choice of location. In certain portions of the State a worm seriously deteriorates the crop. In the Salmon City

---

¹ No coke.

Valley, however, it has not yet made its appearance, and the apples grown here, upon the talus slopes at the foot of the mountains, and in the higher valleys, are the finest in flavor and the largest in size the writer has ever seen. Berries and vegetables may be grown with great ease in all the valleys of the State. The area for large farming operations, the growing of hay, grain, etc., is at present small, but is capable of great increase by the irrigation of the many thousand acres that can be selected on the plains of the Snake River. That such lands are suitable for cultivation under irrigation has already been proved by the successful farms and orchards now in operation. Such irrigation, however, can be accomplished only on a large scale, by heavy outlays of capital. Artesian water is possible at many points, the porosity of the strata, both in the Snake Valley and in the intermontane depressions, rendering a large subterranean circulation quite probable. Hot springs abound, remarkable for their number, distribution, and size. They are already utilized at Boise in the municipal economy.

www.ingramcontent.com/pod-product-compliance
Lightning Source LLC
Chambersburg PA
CBHW020229090426
42735CB00010B/1626